Nanocoatings Nanosystems Nanotechnologies

AUTHORED BY

Alexander D. Pogrebnjak

Sumy State University
Sumy Institute for Surface Modification, R.-Korsakov Str.
2, 40007 Sumy
Ukraine

&

Vyacheslav M. Beresnev

Kharkov National University V.M. Karazina, sq. Svobody
4, 61022 Kharkov
Ukraine

CONTENTS

Introduction — i

About the Authors — ii

Foreword — iii

Preface — iv

CHAPTERS

1. Structural Features of Nanocrystalline Materials — 3

2. Nanoporous Materials — 15

3. Amorphous Materials — 21

4. Fulerene, Fulerite and Nanotubes — 31

5. Nanocomposite Material — 39

6. Methods Employed for Nanomaterial Fabrication — 47

7. Methods of Nanomaterials Investigation — 84

8. Structure and Properties of Nanostructured Films and Coatings — 95

9. Application of Nanomaterials in Engineering — 122

Terms — 143

Index — 145

INTRODUCTION

NANOCOATINGS, NANOSYSTEMS, NANOTECHNOLOGIES

This Reference eBook deals with an existing classification of a nanosized structure and an analysis of its properties. It summarizes an information about how a grain size affects physical, mechanical, thermal, and other properties of a nanostructured material. A basic method, which is employed for a fabrication of an isolated nanoparticle, an ultradisperse powder, a compact nanocystalline, nanoporous, and amorphous material, a fullerene, a nanotube, and a nanostuctured coating, is considered. Investigation methods, which are applied to study the nanostructured material, are briefly described. A modern understanding of a formation of the nanostructured and nanocomposite coating, which are fabricated using a ion-plasma deposition method, is reposted. A potential application of the nanostructured material and coating in a field of engineering is demonstrated.

Readership: Graduate, Postgraduate, Ph.D. Students, Researchers and Industry Professionals.

The eBook contains 9 Chapters, 87 Figures, 14 Tables, 411 References-totally, 155 Pages. It was approved by two Scientific Boards from National Kharkov University and Sumy State University.

A main content of this eBook is a basis for lectures presented for students at Sumy State University (the Physical-Technical Faculty), Kharkov National University (the Physical Faculty), Omsk State University (the Physical Faculty, the Department of Material Science), East-Kazakhstan State Technical University (Ust-Kamenogorsk, Kazakhstan), and Moscow State University (the Physical Faculty, Moscow, Russia).

A source of this eBook is original papers of leading world-known scientists, who ware involved in a field of new nano composite material fabrication, nanotechnologies, and researches. This version had not been published elsewhere. It is interesting for a wide circle of specialists, Masters, aspirants, scientific researchers, and a technical staff of Higher Education System, Research Institutes and Laboratories. It covers recent data since 2008 to 2010 year.

About the Authors

Pogrebnjak Alexander Dmitrievich-Director Sumy Institute for Surface Modification, Sumy State University. Professor, Dr. Sc. 227 referred papers, 52 Author Certificates, 15 patent, 14 review papers, 7 Books. Member of JEC/PISE since 2006. Regular Member of Organizing Committee of 7 International Conferences: AEPSE, PSE, ION, NEET, *etc.* Member of Scientific Board for Ph.D and Dr.Sc. Sc since 1993.

Field of interests: material modification using ion, electron, and plasma beams, ion implantation, deposition of protective and nanocomposite coatings, electrolyte-plasma treatment of steels, micro-arc oxidation and combined coatings, duplex methods of treatment.

Beresnev Vyacheslav Mikhailovich-Professor, Dr. Sc., Professor of Kharkov National University of Ministry of Science and Education. 59 referred papers, 25 patents and Author Certificates, 3 Books.

Field of interests: nanocomposite and nanocrystalline films and coatings, ion implantation, and combined technologies.

FOREWORD

This manual is as a course of lectures for students of physical, technical, and engineering faculties, whose future field of interest is Fabrication of Nanomaterials, Nanosystems, and Nanocomposites as well as Studies of Their Properties". The Manual also deals with relating Methods of Analysis, which are employed for Researches of above mentioned materials, as well as considers Original Works of Leading Specialists Working in the Field of Nanomaterials".

The manual is composed of 9 Chapters. The 1st Chapter reports a general information about nanosized structures, describes mechanical, thermodynamic, electrical, and magnetic properties, and presents the structure classification. The 2nd Chapter is devoted to nanoporous materials, and presents their modern classification. The 3rd Chapter briefly describes fundamentals of amorphous material fabrication and their properties with the purpose of their comparison with nanostructured materials. The 4th Chapter considers fullerenes, fullerites, and nanotubes, as well as briefly describes their definitions and properties. The 5th Chapter briefly describes nanocomposite materials, methods of their fabrication, and properties. The 6th Chapter describes methods of nanomaterial fabrication and their classification as well as methods of nanomaterial and amorphous material fabrication, which are employed in powder metallurgy. Methods of an intensive plastic deformation, which are currently employed for a material fabrication, as well as CVD and PVD methods, which are employed for thin film fabrication, are considered. In addition, the Chapter deals with methods employed for fullerene and nanotube fabrication. It describes a particle micro-and nanoprobe applied for an analysis of the nanosized material. The Chapter 7 presents methods applied for nanomaterial researches, such as structure and chemical analysis, a mechanical testing of solids, for example, measurements of nanohardness, an elastic modulus, *etc.* Chapter 8 considers structure and properties of a nanostructured film and coating. This Chapter is the biggest one in its volume. It considers many problems, such as role of energy in a coating formation, effect of an ion bombardment on the formation of the nanocrystalline coating. It presents information about a mixing process, the formation of a multi-layered coating, fabrication of super-hard coatings, thermal stability, and oxidation resistance of a solid coating, as well as describes their mechanical properties. The Chapter reports why temperature and potential applied to a substrate play a crucial role in the formation of the nanocrystalline coating.

The manual is written by specialists involved for many years in researches of a material modification by particle beam and plasma, problem of new nanostructured material and combined coating fabrication, as well as all aspects of an ion implantation. The content of this eBook is well written and expressed in a compact and intelligible form. This manual is interesting not only for senior students, PhD students, but also for researchers, physicists, and chemists, whose field of interest is material science and solid, as well as for various Institutions such as Universities, Institutes, and Enterprises.

Prof. F. F. Komarov
Corresponding Member of the National Academy of Sciences of Belarus
UK

PREFACE

Alexander D. Pogrebnjak[1] and Vyacheslav M. Beresnev[2]

[1]*Sumy State University, Sumy Institute for Surface Modification, R.-Korsakov Str., 2, 40007 Sumy, Ukraine; E-mail: alexp@i.ua and* [2]*Kharkov National University V.M. Karazina, sq. Svobody, 4, 61022 Kharkov, Ukraine; E-mail:Beresnev-scpt@yandex.ru*

In Recent years, a research of materials, which are composed of submicron nanosized grains and clusters, are swiftly developing due to already existing and/or potential applications in many technological fields such as electronics, catalysis, magnetic data storage, structure components, *etc.*

Metallic and ceramic materials with a submicron and nanocrystalline grain structure are now widely used as construction elements and functional layers in a modern microelectronics, as sites of devices in aviation and space engineering, and as hard wear resistant coatings in industry. To satisfy the technological requirements of these industrial fields, the size of structure elements is to be decreased to a submicron and a nanometer range. However, when a size of structure element decreases to a nanometer range, a material starts to demonstrate radically new physical and mechanical properties in comparison with a bulky base. Researches of these *nanosized structures* (nanostructures) rank among nanotechnological directions. Development and researches of nanostructured materials (further referred to as nanomaterials) and nanostructure properties obtained under various conditions are very important components of these scientific-technological directions. A material, a structure of which is composed of grains of about 0.3 to 0.04 μm size, is considered as a submicrocrystalline [1-3]. A material of smaller grain size is considered as a nanomaterial.

A nanomaterial (a nanocrystal, a nanocomposite, a material with a nanophase structure, *etc.*) is to be understood as a material, in which structure elements (a grain, a crystallite, a fiber, a layer, a pore) do not exceed a limit of 100 nm (1 nm = 10^{-9} m), at least, along one crystallographic direction. According to size of a structural unit, the nanomaterial is conventionally subdivided into a nanocluster and nanocrystalline material. A nanocluster material is subdivided into small (3 to 12 number of atoms, 100% of surface atoms, without an inside layer), big (13 to 150 number of atoms, 92 to 63% of surface atoms, including 1 to 3 inner layers), and giant nanocluster material (151 to 22000 number of atoms, 63 to 15% of surface atoms, including 4 to 18 inside layers). Conventionally, a cluster top boundary corresponds to such amount of atoms that an addition of one more atom already cannot change physical-chemical properties of this cluster. Theoretical calculations, which were confirmed by experimental researches for a cluster containing not less than 300 atoms, demonstrated that an icosahedrons structure is the most stable one. When an amount of cluster atoms increased, an elastic deformation energy quickly rouse in a proportion to their volume, and consequently, this icosahedrons structure is destabilized forming a face-centered cubic lattice [4].

A structure unit with a higher amount of atoms and 3 to 40 nm grain size ranks among a nanocrystal. This nanocrystalline material has various forms and demonstrates unique chemical, physical, and mechanical properties. A grain size is limited by the maximum size of the nanostructure elements and depends on some critical parameters (a size effect): a free range length of carriers participating in an energy transfer, a size of a domain/a domain wall, a diameter of a Frank-Reed loop, a de Broglie wave length, *etc.* This size effect sharply changes quality and properties of the nanostructured system and indicates a special condensed material state, which exists only in the nanostructured material. Today, the nanostructured material can be formed on the basis of various metals and alloys, and with the help of specially developed technological methods.

In recent years, a definite progress had been achieved in physical researches and technologies of the nanostructured material fabrication. In particular, an important stage of these researches is a systematic study of microprocesses occurring in a phase interface in the course of nanostructured system formation. This systematic study stimulated an appearance of calculation methods, which are employed to predict optimal technological parameters and promising ways of the nanostructured material formation.

A whole number of publications, monographs, and papers [5-11] report about technologies, structures, properties, and applications of the nanomaterial and the nanostructure.

Here, we present only a description of individual representatives and classes and do not reflect, to a full extent, features of this modern direction. Why is there this modern interest in a nanotechnology, in general, and in a nanostructure study, in particular?

On one hand, nanotechnologies allow formation of a principally new material, which can find its application in future, since it is compact and functionable. It plays an important role in the formation of principally new elements for future nanodevices, which are dependent on physical principles employed for their functioning.

On the other hand, the nanotechnology is an extremely wide interdisciplinary direction, uniting specialists working in a field of physics, chemistry, materials science, biology, technology, directions of intellectual/self-organized systems, high-technological computer engineering, *etc.* Finally, solving problems arising in the field of nanotechnologies, and, first of all, in the process of researches, scientists find many gaps existing both in fundamental and technological knowledge. All above mentioned excites a concentrated interest of a scientific and engineering society to this direction [12-21].

In many technologically advanced countries such as USA, United Kingdom, Japan, China, Russia, national programs, which are specified at an intensive development of various directions of the nanotechnology and formation of new nanostructures, are accepted and have started to be actively introduced into a practice.

Now, several basic types of the nanomaterials are known [1, 4].

VARIETY OF NANOMATERIALS

A nanomaterial has a number of structure characteristic features, which are the parameters relating to a structure as a whole and those identifying its individual elements. In their turn, the structure characteristic features of the nanomaterials are reflected in an unusual display of their properties. Since the nanomaterial is a basic unit of a nanosystem, properties of the nanosystem to a considerable degree depend on the nanomaterial properties.

Variety of nanomaterials is immense and every type is characterized by a specific structure and, as a consequence, specific properties. The characteristic features of the nanomaterial and the system formed on its basis, first of all are manifested in a size effect, among which a quantum effect takes a special place.

According to degree of their structure complexity, the variety of nanomaterials is subdivided into materials composed of individual nanoparticles and those composed of nanostructures (Fig. **1**).

A nanoparticle is a nanosized complex of atoms and molecules, which are interrelated in a definite way.

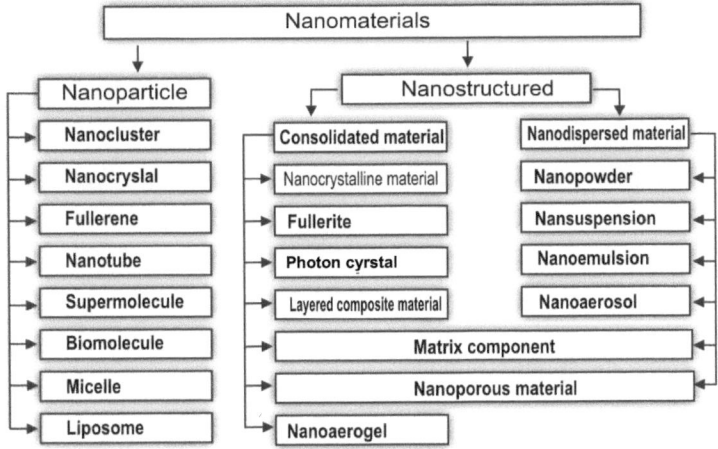

Figure 1: A classification of nanomaterials according to their structure characteristic features.

The following types of the nanoparticles are identified:

- Nanocluster, which is sorted as an ordered cluster characterized by a definite order in an arrangement of atoms and molecules and a strong chemical bond and a non-ordered nanocluster characterized by a disordered arrangement of atoms and molecules and a weak chemical bond;

- Nanocrystal (a crystalline nanoparticle), characterized by the ordered arrangement of atoms and molecules and the strong chemical bond like a bulky crystal (a macrocrystal);

- Fullerene, which is composed of carbon atoms (or atoms of another element) forming a structure looking like a spherical carcass;

- Nanotube, which is composed of carbon atoms (or atoms of another element) forming a structure looking like a cylindrical carcass closed at its both ends;

- Supermolecule, which is composed of "a host molecule" with a three-dimensional structure, in a cavity of which a "guest molecule" is arranged;

- Biomolecule, which is a complicated molecule of biological origin characterized by a polymer structure(DNA, a protein);

- Micelle, which is composed of molecules of a surface-active matter forming a sphere-like structure;

- Liposome, which are composed of molecules of a special organic compound like a phospholipid forming a spherical structure;

A nanostructured material is an ensemble of nanoparticles. Nanoparticles play a role of a structure element in such material. A type of the nanostructured material depends on a character of interrelation existing between nanoparticles: a consolidated material and a nanodispersed one.

The consolidated material is a compact solid-phase material, which is composed of nanoparticles with a fixed spatial position in the material volume and rigidly-directly bound to another one.

The consolidated material is:

- Nanocrystalline material, which is composed of nanocrystals usually called a nanograin or a nanocrystallite;

- Fullerite, which is composed of fullerenes;

- Photon crystal, which is composed of ordered-in-space elements, a size of which is comparable with a half-length of a photon wave in one, two or three directions;

- Layered composite material (with a superlattice), which is composed of various material layers of a nanosize thickness;

- Matrix component, which is composed of a solid base (a matrix), in the volume of which nanoparticles (nanowires) are distributed;

- Nanoporous material, which is characterized by the presence of nanopores;

- Nanoaerogel, which is composed of an interlayer of a nanosize thickness separating pores.

A nanodispersed material is a dispersed system with a nanosized dispersion phase.

In addition to the above mentioned matrix nanocomposite materials and nanoporous materials, the nanodispersed materials cover:

- Nanopowder, which is composed of contacting other nanoparticles;

- Nanosuspension, which is composed of nanoparticles free-distributed in the liquid volume;

- Nanoemulsion, which is composed of nanodrops of a liquid free-distributed in a volume of another liquid;

- Nanoaerosol, which is composed of nanoparticles and nanodrops free-distributed in a volume of a gaseous medium.

Specimens of various nanostructured materials are often bulky, *i.e.* are characterized by a micro-and macro-size, whereas their structure elements are nanosized.

Effects, which are related to the small size of composing structures, may manifest themselves in a different way in various nanomaterials.

For example, a specific surface of the nanocrystalline and nanoporous material is crucially larger, *i.e.* a fraction of atoms arranged in a thin (about 1nm) near-surface layer radically arises. This increases the reaction ability of the nanocrystal, since atoms, which are arranged in the surface, have unsaturated bonds in contrast to atoms, which are arranged in the material bulk, since they are bound with surrounding atoms. A change in atomic ratio between the surface and the bulk atoms may result in atomic reconstruction, in particular, in a change of an atomic arrangement order, an interatomic distance, and a crystalline lattice period. The size dependence of a nanocrystalline surface energy predetermines a corresponding dependence of a melting temperature, which is lower for the nanocrystal than for the macrocrystal. As a whole, heat properties of the nanocrystal are crucially different, which is related to a character of atomic heat oscillations.

When the size of ferromagnetic particle decreases to a certain critical value, the domain separation becomes energetically disadvantageous. As a result, polidomain nanoparticles become single domain and acquire special magnetic properties, which are manifested in supermagnetism.

The fullerene and the nanotube are characterized by very unusual properties due to their specific structure. This is true also for the molecular and biomolecular complex functioning according to laws of molecular chemistry and biology.

Peculiarities of a structure and properties of an individual nanoparticle affect in a definite way a structure and properties of the consolidated materials and the nanodispersion, which are formed on their basis.

A typical example is a nanocrystalline material, which is characterized by a decreased grain fraction, and, respectively, an increased fraction of interfaces occurring in the material volume. Simultaneously, a change of structure characteristics of both grains and interfaces takes place. As a result, mechanical properties of the nanocrystalline material significantly change. The material demonstrates a superhardness of a superplasticity under definite conditions.

Electron properties of the nanostructure, which are conditioned by quantum effects, are of a special interest for practical applications.

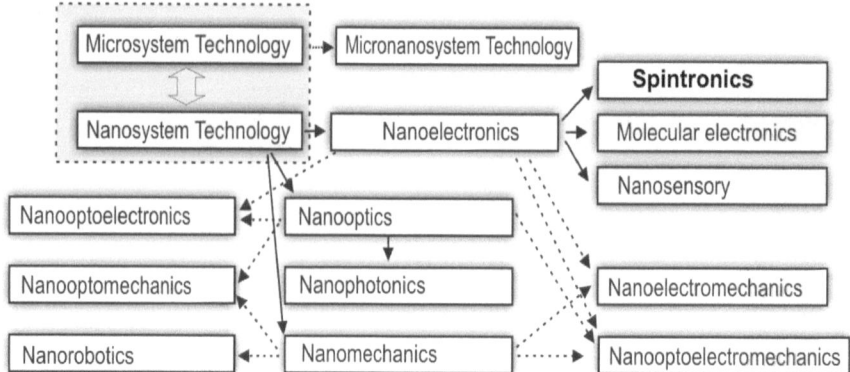

Figure 2: A classification of types of nanosystem-based devices according to their functional purposes.

The nanomaterial serves as a basis for the development of nanosystems with various functional purposes, which in their turn are subdivided into an electron, an optical, and a mechanical nanosystem, according to the principle of functioning, Fig. **2**. An action of the electron nanosystem is based on transformation of an electrical signal, that of the optical one-on transformation of an optical signal (light) into the electrical one and *vice versa*, and the mechanical nanosystem transforms a mechanical motion.

Sets of definite characteristics of nanosystems are employed in definite fields of engineering such as nanoelectronics, nanooptics, and nanomechanics. Development of various types of nanosystems is closely interconnected and results in fabrication of more constructively complicated and integrated nanosystems, such as nanooptical-electron, nanoelectrical-mechanical, nano optical-mechanical, and nano optical-electro-mechanical systems.

The development of nanosystems is undoubtedly a new step, which will enable a future progress of microsystems. In practice, nanosystems are built-in various Microsystems forming in this way a promising direction of a modern system units (devices) such as a micronanosystem equipment.

A consolidated material-is a compact, a film, or a coating formed from a metal, an alloy, or a compound using a powder technology, an intensive plastic deformation, a controlled crystallization from an amorphous state, and various other techniques, which are currently applied for deposition of a film and a coating.

A nanosemiconductor, a nanopolymere, and a nanobiomaterial may exist in an isolated or a partially mixed (consolidated) state.

Fullerene and nanotube became an object of researches since the moment, when Sir Harold (Harry) Walter Kroto (1985) found a new carbon allotrope form-a cluster C_{60} and C_{70}, which was called fullerene. This new carbon form attracted much more attention, when the carbon nanotube was revealed in a graphite product after an electrical-arc evaporation (Sumio Iijima, 1991).

Nanoparticles and nanopowder represent a quasizero grain size structure having various compositions and a size of which usually does not exceed a nanotechnological limit. A difference is that the nanoparticle is isolated, while the nanopowder is an aggregate. In a similar way, the nanoporous material is characterized by a pore size, which, as a rule, is not less than 100 nm.

A supermolecular structure is a structure, which is formed as a result of so-called non-covalent synthesis accompanied by formation of weak bonds (a Van der Waals, a hydrogen type, *etc.*) between molecules and their ensembles.

The nanomaterial is not a "universal" material; it is a vast class of many various materials joining different families. In addition, there exists a delusion that the nanomaterial is a material composed of very small "nano"-particles. In reality, many nanomaterials are not composed of individual particles; they are complicated micro objects nanostructured either in a surface or in a volume. Such nanomaterials are considered as a special state of matter, since properties of these materials, which are composed of nanostructured and nanosized elements, are not identical to properties of a volume material.

So, the nanomaterial is characterized by several basic features, which position it beyond any competition in comparison with other matters.

First, the nanomaterial is composed of very small objects, which cannot be seen with a naked eye. It represents a "super miniaturization", which leads to a possibility that more and more quantity of functional nanodevices can be placed at an area unit. This is vitally important, say, for nanoelectronics or very dense magnetic information recording, which can reach 10 Tirrabit per a square centimeter.

Second, the nanomaterial has a large surface area, which promotes a hastened interaction inside and within a medium, to which it is placed in. For example, a catalytically active material can hasten a chemical and biochemical reaction by a factor of ten, thousand, and even a million [22-30].

Decomposition of water into hydrogen and oxygen for needs of a hydrogen power engineering, which is realized in the presence of titanium dioxide nanoparticles (everyone knows it as a component of titanium white paint), seems to be very interesting. A nanofilter can screen bacteria or efficiently absorb impurities and toxins.

Third, the nanomaterial has unique physical and mechanical properties, and this means that such a matter is in a specific "nanosized" state. Changes found in the nanomaterial fundamental characteristics are conditioned not only by a small grain size, but also by a quantum-mechanical effect, in which an interface plays a dominating role. The effect arises when the grain size is so "critical" that is commensurable to a so-called correlation radius or other physical parameters (for example, a free electron and a phonon range, a coherence length in superconductors, a magnetic domain or a nucleon size of solid phase, *etc.*). This makes, in particular, a semiconducting material to be an ideal element for a perfect energy-consuming laser and light emission. Hardness of an individual carbon nanotube exceeds that of the best steel by a factor of ten. At the same time, it has many-fold advantage in a specific mass. All above-mentioned characteristics fully explain the fact that even a gram of the nanomaterial may be much more efficient than a ton of an ordinary matter, and that its industrial production is not a problem of quantity, ton, and kilometer, but that of a human thought quality, *i.e.* "know-how".

Nanotechnology is an extremely complicated, professional, interdisciplinary field, which needs joined efforts of chemists, physicists, specialists in material science, mathematicians, medics, specialists in calculation methods, *etc.* Deep scientific fundamentals are admirably interwoven in a field of nanomaterials with aspects of a human knowledge and practical applications.

In this eBook, we report fundamental data concerning structure, properties, and application of the modern nanomaterials. In the First Chapter, we present general information about nanomaterials, their structure features, size effect on structure formation and on physical-mechanical properties.

In Chapters 2 and 3, we present information about structure and properties of a nanoporous and an amorphous material.

In Chapter 4, we consider certain properties of fullerene and nanotube. Chapter 5 deals with a nanocomposite based on a polymer. Chapter 6 is devoted to methods, which are currently employed for the nanomaterial fabrication, since these new methods really gave rise to a violent development of this field.

Physical research methods, namely, novel methods employed for surface studies are presented in Chapter 7.

In Chapter 8, we consider mechanical and thermal properties of a nanocrystalline film and a nanocomposite coating, which are fabricated using physical deposition methods.

Chapter 9 is devoted to an application of the nanocrystalline material, which is employed in an engineering society.

REFERENCES

[1] Morokhov ID, Trusov LI, Chizhij SP. Unltradipersion Metallic Media. M Atomizdat 1977.
[2] Gleiter H. Nanostructured Materials. Basic Concepts and Microstructure. Acta Materialia 2000; 48: 1-29.
[3] Seigel RW. Nanostructured Materials-Mind over Matter nanostruct. Mater 1993; 3: 1-18.
[4] Larikov LN. Nanocrystalline Compounds of Metals Metallo-Fizika I Noveishie Tekhnologii 1995; 17: 56-68.
[5] Andrievskii RA, Ragulia AV. Nanostructured Materials. Moscow: Academia 2005.
[6] Roko MK, Wiliyams PS, Alivisatos P. Nanotechnologies in the Next Ten years. Prediction of Researching Directions; Translation. Ed. Andrievskii RA. Moscow: Mir 2002.
[7] Liakishev NP, Alymov MI. Nanomaterials for Structural Materials. Nanotechnologies in Russia 2006; 1: 71-81.
[8] Gusev AI. Nanomaterials, Nanostructures, Nanotechnologies. Moscow: Fizmatlit 2005.
[9] Gusev AI, Rempel AA. Nanocrystalline materials. Moscow: Fizmatlit 2000.
[10] Pozdniakov VA. Physical Material Science for Nanostructured Materials. Moscow: MGIU 2007.
[11] Ragulia AV, Skorokhod VV. Consolidated Nanostructured Materials. Kiev: Naukova Dumka 2007.
[12] Sergeev TB. Nanochemistry Moscow: MGU 2003.
[13] Andrievskii RA, Glezer AM. Size Effects in Nanocrystalline materials. I. Features of Structure. Thermal-Dynamics. Phase Equilibrium. Kinetic Phenomena. The Physics of Metals and Metallography 1999; 88: 50-73.
[14] Andrievskii RA, Glezer AM. Size Effects in Nanocrystalline Materials. II. Mechanical and Physical Properties. The Physics of Metals and Metallography 2000; 89: 91-112.
[15] Pogrebnjak AD, Shpak AP, Azarenkov NA, Beresnev VM. Structure and Properties of Hard and Superhard Nanocomposite Coatings. Physics Uspekhi. 2009; 52(1): 29-54.
[16] Noskova NI, Muliulukova RR. Submicrocrystalline and Nanocrystalline Metals and Alloys. Ekaterinburg: Ural'skoe Otdelenie RAN 2003.
[17] Suzdalev IP. Nanotechnology: Physical Chemistry of Nanoclusters, Nanostructures, and Nanomaterials. Moscow: KomKniga 2006.
[18] Valiev RZ, Aleksandrov IV. Nanostructured Materials, Fabricated by Intensive Plastic Deformation. Moscow: Logos 2000.
[19] Andrievskii RA. Nanomaterials: Concepts and Modern Problems. Rus Chem J 2002; XLVI: 50-56.
[20] Glezer AM. Amorphous and Nanocrystalline Structures: Similiarities, Difference, Mutual Transitions. Rus Chem J 2002; XLVI: 57-63.
[21] Andrievskii RA. Thermal Stability of Nanomaterials. Russ Chem Rev 2002; 71: 967-981.
[22] Demikhovskii VJ, Vugalter GA. Physics of Quantum Small Grain Size Structures. Moscow: Logos 2000.
[23] Harris P. Carbon Nanotubes and Related Structures. New Materials of XXI Century. Translated by L.A.Chenazatonskii. Moscow: Tekhnosfera 2003.
[24] Pomogailo AD, Pozenberg AS, Ufliand IE. Nanoparticles of Metals in Polymers. Moscow: Khimiia 2000.
[25] Shevchenko SV, Stetsenko NN. Nanostructured States in Metals, Alloys, and Intermetalloid Compounds: Methods of Fabrication, Structure, Properties.Progress in Physics of Metals 2004; 5: 219-255.
[26] Shik AJ, Bakuleva LG, Musikhin SR, Rozhkov SA. Physics of Small Grain Size Systems St.-Peterburg: Nauka 2001.
[27] White Book on Nanotechnologies: Researches in the Field of Nanoparticles, nanostructures, and Nanocomposites in Russian Federation (Materials of the First All-Russia Meeting of Scientists, Engineers, andfabricationrs in the Field of Nanotechnologies) Moscow: LKI 2008.
[28] Pul Ch, Ouens F. Nanotechnologies. Translation by Moscow: Tekhnosfera 2004.
[29] Golovin YuI. Introduction in nanotechnology Moscow: Mashinostroenie 2003.
[30] Shorshorov MKh. Ultradispersion Structure State of metallic Alloys. Moscow: Nauka 2001.

2

CHAPTER 1

Structural Features of Nanocrystalline Materials

Abstract: In this Chapter, general information about a nanosized structure is presented, as well as terms and definitions are introduced. Effect of the nanostructure on mechanical, thermo-dynamical, electrical, and magnetic properties of materials is described. Questions for self-control are presented at the end of this Chapter.

Keywords: Classification, properties, nanostructured materials, interface.

1.1. GENERAL INFORMATION ABOUT NANOSIZED STRUCTURE

Recently, an interest to researches of materials with a nanocrystalline structure grew higher, since it was found that a decrease of a crystallite size (or any other structure element) below some threshold value resulted in a radical change of material physical properties. On one hand, a tendency to a further miniaturization of devices, which are employed in microelectronics, is a beginning of a new spire in researches. On the other hand, this spire is stimulated by works, which appear in the middle of 80'S, and present a classification of such materials [1-8]. A great contribution of Prof. Gleiter, who is the author of these works, is not only an assumption that a big class of different materials (such as an ultradispersed, a composite, a granulated material, powder, *etc.*) is the same class of nanostructured materials, united by one common property-a size of their structure elements. Prof. Gleiter also reveals that they own some characteristic special properties [9]. Both these features and the nanocrystallite size are a reason of their different physical properties. According to a chemical composition and phase distribution, four structure types are distinguished (Table 1): a single-phase structure, a static many-phase structure with identical and non-identical interface surface, and a matrix multi-phase structure.

Table 1: A classification of a consolidated nanomaterial.

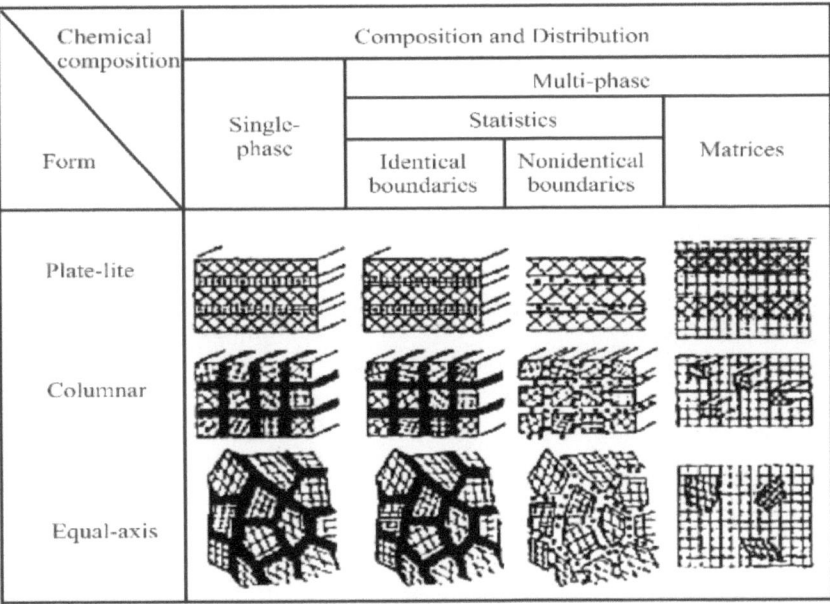

According to a form, three structure types are distinguished : a plated, a columnar, and a form containing equal-axes inclusions. Here, a possible segregation at a crystalline interface is also taken into account. A single-and multi-phase matrix, a static object, a columnar, and a multi-layered structure (a film) are most popular.

Here, a possible segregation at a crystalline interface is taken into account.

Alexander D. Pogrebnjak and Vyacheslav M. Beresnev
All rights reserved-© 2012 Bentham Science Publishers

In general, an abundance of interface surfaces (a grain interface and a triple-junction, which are a junction of three grains) characterizes the nanomaterial structure. Schematically, this triple-junction looks like a tetrahedral dodecahedral structure (Fig. **1a**).

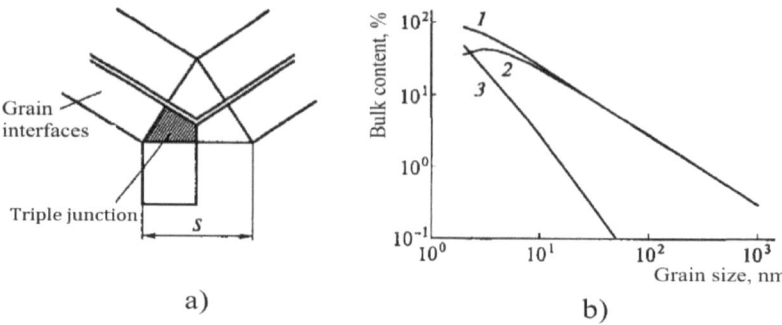

Figure 1: A scheme of a triple-junction *(a)* and a grain size effect *(b)* on a total interface fraction *(1)*, a grain interface *(2)*, and a triple-junction *(3)* when $s = 1$ nm.

According to Fig. **2a**, a total fraction of interfaces is

$$V_{n.p.} = 1-[(L-s)/L]^3 \sim 3s/L \tag{1}$$

A fraction of grain interfaces *per se* is

$$V_{m.r.} = [3s(L-s)^2]/L^3, \tag{2}$$

And, correspondingly, a fraction of triple-junctions

$$V_{t.c.} = V_{n.p.} - V_{m.r.}, \tag{3}$$

where: L is a grain size, s is an interface width (a zone near the interface).

Fig. **1b** presents how a total fraction of interface surfaces, grain interfaces, and triple-junctions depend on a grain size. It is seen, that an essential fraction of interface surfaces (several per cent) corresponds to a grain size $L < 100$ nm. Thus, even when $L < 10$ nm, the fraction $V_{n.p.}$ is about several tens of percent. Within this interval of L values, the fraction of triple-junctions quickly increases. The fraction of interfaces $(V_{n.p.})$ over the whole material volume is approximately equal to $3s/L$. So, for example, for $s = 1nm$ and $L \approx 6$ nm, the fraction $V_{n.p.}$ already amounts 50% [9-18].

Relaxation processes, which are inevitable under thermal action and in a power field, should affect physical-chemical, physical-mechanical, and other properties and influence also servicing resources of the nanomaterial. Therefore, studies of their stability are very important.

Fig 2. presents an effect of the grain structure state (the size effect) on nanomaterial properties.

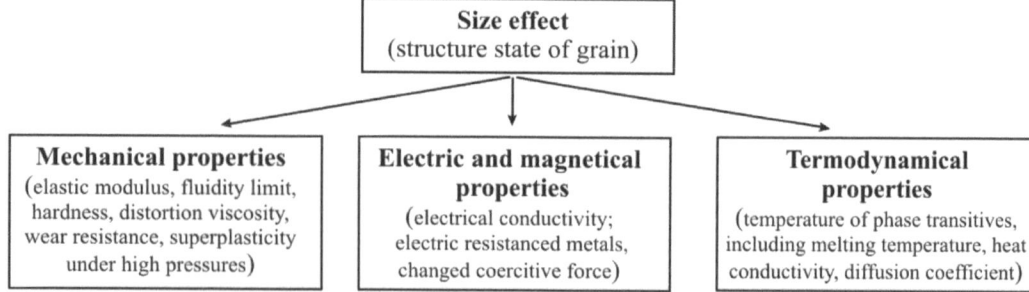

Figure 2: An effect of a grain structure state on nanomaterial physical-mechanical properties.

The nanostructure features the following properties:

- When an element size decreases, a role of the interface surface increases (a fraction of near surface atoms increases from one percent to several scores of percent);

- Properties of a material with the interface surface ranging within a nanometer interval may differ essentially from a big-sized crystal (an edge effect);

- Size of the nanostructure element is commensurable with a range of a certain physical phenomenon (for example, with a free pass length of carriers under transfer);

- Thermodynamically "profitable" self-organization and a self-construction, which decrease a redundant free energy of a system;

- An effect of the nanostructure size may have a quantum character (when a localization area of free carriers becomes commensurable with L, a de Broglie wave length λB (Fig. 3) is

$$\lambda B = \frac{2\pi h}{\sqrt{2mE}} \tag{4}$$

where m is an efficient electron mass; E is a carrier energy; $\hbar$ is Plank constant.

A square-law dependence of an electron state density $N(E)$ on an energy is characteristic for a microscopic structure. A decrease of the carrier localization areas up to λB in one, two, or three directions, as it follows from a solution of a Schrödinger equation under corresponding boundary conditions, is accompanied by a change of $N(E)$ dependence character.

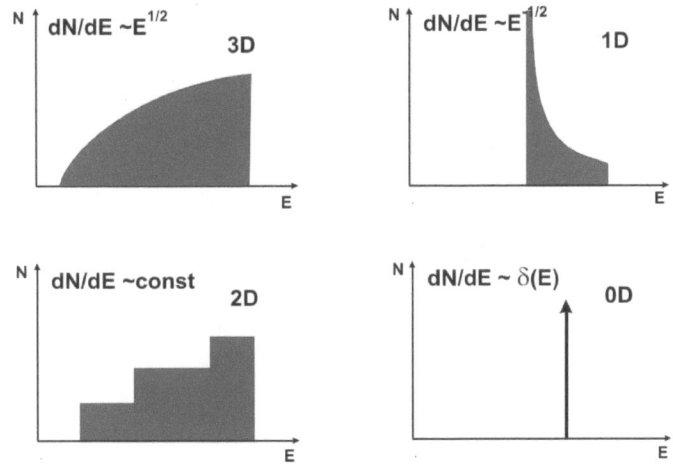

Figure 3: A charge carrier state density $N(E)$ for structures of various grain sizes.

A free motion of charge carriers in two-and one-dimensional structures is two-and one-dimensional, respectively. An electron energy spectrum at quantum points is quantized in three dimensions and represents a set of discrete levels, which are separated by zones of a forbidden state [19-28].

Additionally, when the size of the nanostructure element decreased:

- Width of the forbidden zone increased;

- An absorption band shifted towards high energies, according to theoretical calculations $E \sim 1/R$;

- "Blue" shifting took place (*i.e.* a shift of a luminescence spectrum towards a short-wave band);

- Dependence of electrical properties on the size became non-monotonous inducing a modification of nanostructure electron properties.

Carbon tubes are different in their forms, electron structures, and properties. An important characteristic of the nanotube structure is a chirality, which is a mutual orientation of hexagonal graphite net with respect to the nanotube longitudinal axis. A conductivity type of a single-layered nanotube depends on the chirality: for example, a zigzag tube has a metallic-type conductivity and a chirality tube has a semiconductor one. The forbidden zone width correlates to a tube radius (for a *p*-type semiconductor $\Delta E_g \sim R^{-2}$, for an *n*-type semiconductor $\Delta E_g \sim R^{-1}$).

Fig. **4** shows the nanocomposite material consisted of unlikely-charged crystallites: a) a *p*-crystallite (the *p*-type semiconductor) and an *n*-crystallite (the *n*-type semiconductor); b) phases of a different Fermi energy; c) a metal and a semiconductor. An application of an external electromagnetic field changes a charge of the interface surface.

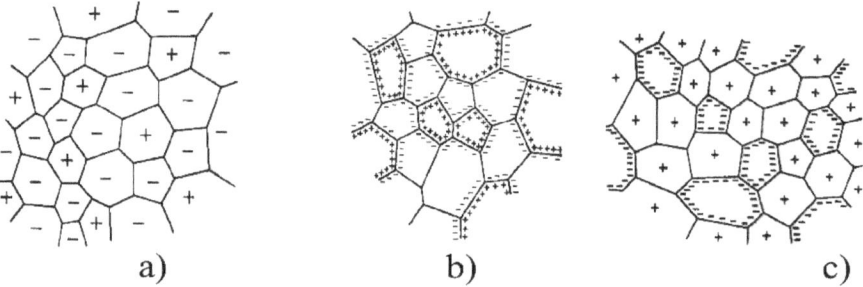

Figure 4: A nanocomposite structure containing phases of different charges [15].

1.2. MECHANICAL PROPERTIES

Properties of the nanocrystalline structure differ from bulky analogs. In particular, an extraordinary hardness seems to be the first among mechanical characteritics. Since hardness characterizes a material resistance to plastic deformation under diamond indentation, it is natural to assume a proportional relation of hardness to a material fluidity limit (σy).

Historically, namely, the material fluidity limit is the first parameter, which is subjected to a detailed analysis for its size dependence. As a result, a relation, which is called a Hall-Petch law, is derived:

$$\sigma_y = \sigma_0 + k_y D^{-n}, \tag{5}$$

where: σ_0 is an internal voltage preventing a dislocation motion, k_y is a coefficient relating to a grain interface penetrability for dislocation motion. A value of *n* changes from ½ (a classic Hall-Petch law) to values within ¼ ...... 1 interval. It is accepted that every *n*-value corresponds to a certain mechanism characterizing a dislocation interaction with the grain interface. When the grain size exceeds 10^{-6}m, this classic law is successfully fulfilled: this index is $n = ½$ for a metal and an alloy.

Fig. **5** shows how hardness and a fluidity limit change in the case when the grain size decreases to a critical value of about 10 nm. High-resolution electron microscopy data confirm that this value is critical for a formation of dislocations.

Dislocations are absent in crystallites of $d < 10$ nm, and a phase structure of the grain interfaces is close amorphous.

When the grain size decreases to 7nm and smaller, an inverse Hall-Petch effect is observed instead of the increased hardness, which seems to be a result of the decreasing grain size, *i.e.* a material looses its strength.

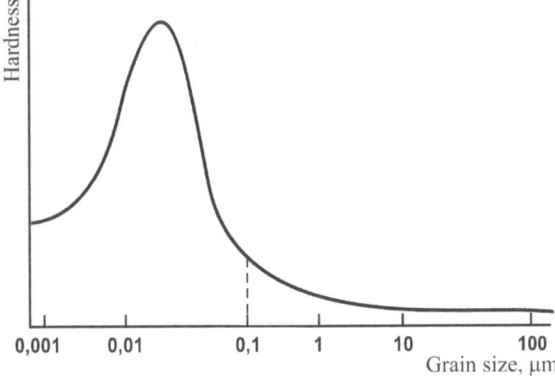

Figure 5: A change of hardness and a fluidity limit with a decreasing grain size.

To explain this anomalous behavior of the nanocrystalline material, the grain size of which is smaller than the critical value, several mechanisms are offered. A sliding along the grain interface (which is sometimes called as a rotation strength loss), a triple junction effect, a presence of nanopores and inclusions, *etc.* are among them.

A change of metal mechanical properties depending on the grain size is presented in Tables **2** and **3** [9-31].

Tension strength of the nanocrystalline material is 1.5 to 2 times higher than that of big-grained analogs. When the grain size decreases below some critical value, hardness decreases due to an increasing amount of triple junctions at the grain interface, which prevents the dislocation motion. The strength increases further with the grain size since a density of already existing dislocations decreases and new ones fail to be formed.

Table 2: Mechanical properties of an ordinary nanocrystalline nickel.

Properties	Microcrystalline Size,10 mkm	Nano-Ni	
		100 nm	10 nm
Strength (hardness), MPa (25 °C)	103	690	>900
Limiting tensile strength, MPa (25°C)	403	1100	>2000
Hardness by Vickers(25°C)	140	300	650

Table 3: A change of mechanical properties depending on a grain size.

Material	Grain Size, μm	T, °C	Strength Limit, MPa	Relative Elongation Till Destruction, %
Titanium alloy BT1-00 (Ti99%, Fe,Si,O)	50	20	380	29
	0,1		730	18
Titanium Alloy Ti-6Al-4V	10	20	1050	9
	0,4		1300	7
	10	600	585	46
	0,4		200	200
Titanium Alloy BT8 (Al6%, Mo3,5%, Si0.4%,Ti)	5	20	1050	45
	0,06		1400	53
Alloy of Ni RSR Rene 80	100	20	375	30
	0,2		850	33

Alloy Al-Mg-LiSc-Zr	10	20	450	5
	0,2		600	6
Steel Fe-25%Cr-0,2%Ti0,12C	50	20	485	26
	0,2		730	17

Fig. **6** presents a strength-to-ductility ratio for a steel.

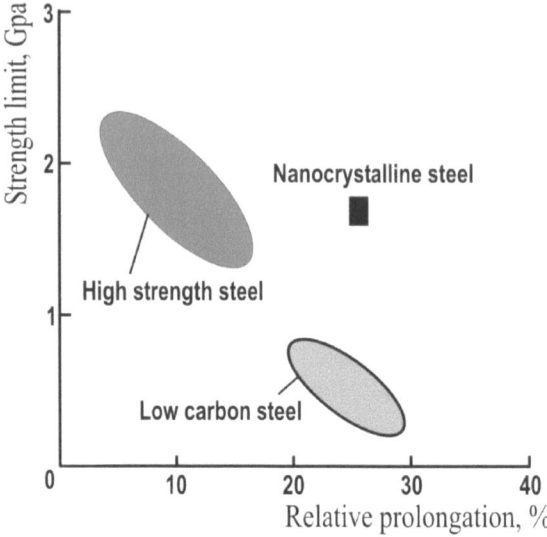

Figure 6: An alternation of mechanical properties for different materials.

A nanocrystalline steel 12X18H10T has a good strength-to-ductility ratio. In some cases, a low ductility of the nanocrystalline material is probably induced by a complexity in a formation of dislocations, in a multiplication, and in a motion, as well as by a porosity, by a presence of microcracks, and inclusions. A wear resistance of aluminum alloy with the nanocrystalline structure is essentially higher that that of big-grained one (Fig. **7**) [29-31].

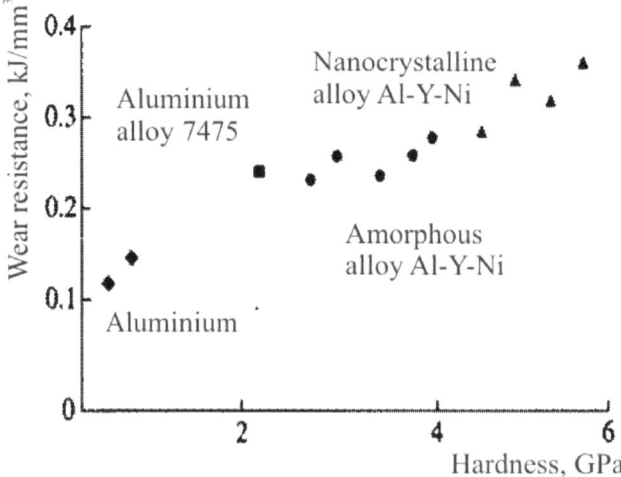

Figure 7: A wear resistance of an aluminum alloy.

Such brittle material as an intermetalloid gained a plasticity when its grain size decreases to some critical value, which is explained by a specific mechanism of multiplication and propagation of microcracks [15-17]. Thus, the nanosize material structure opens a unique chance to reach a new level of properties: high strength, hardness, and wear resistance together with essentially high plasticity.

Knowledge of size regularities open a possibility to realize transition to a new generation of materials, which properties are controlled by a size and a form of their structure elements. These peculiarities may be employed also in units and sites of devices.

1.3. THERMODYNAMICAL PROPERTIES

Speaking about peculiarities of thermal-dynamical properties and phase equilibrium states, we cannot find any unambiguous interpretation. There is an opinion [16] that traditional concepts about surface energy can be more readily applied to particles, a size of which exceeds 10 nm. When the size varies from 1 to 10 nm, properties need some individual specification, and when it is lower than 1nm, a particle gains properties of a whole surface layer and needs a special approach. For the first approximation, these considerations are applicable for a consolidated material thermodynamics [17-19, 27-31].

The following thermodynamical features characterize the nanostructure:

- Certain thermodynamical regularities are not fulfilled since an additional degree of freedom arises due to appearance of identical and independent small-size particles (systems);

- A fluctuation of thermodynamical variables becomes commensurable to an average value of properties themselves;

- A considerable heat effect starts to manifest itself ΔH;

- A phase transformation temperature changes;

- Certain thermodynamical properties (for example, a free Gibbs energy, G) may loose a monotonous character.

As it follows from general considerations, a great number of near surface atoms seems have a significant effect on a background spectrum and related to it thermal properties (a heat capacity, a thermal expansion, a melting temperature, lattice components of a heat conductivity, *etc.*). Additional low-and high-frequency mode is found in nanostructure background spectra. Practically in all cases, the heat capacity increases (under $T < 1K$) in a non-monotonous way, the characteristic temperature decreases, a factor reflecting an atomic shift (Debye-Waller factor) increases, and the melting temperature decreases.

1.4. ELECTRIC PROPERTIES

It is known, that an electrical resistance of a metallic solid is determined mainly by an electron scattering at phonons, structure defects, and impurities. It is noted that a specific electrical resistance ρ increases in many metal-like nanomaterials (Cu, Pd, Fe, Ni, Ni-P, Fe-Cu-Si-B, NiAl, nitrides, borides of transition metals, *etc.*) when the size of structure element decreases. It is due to an increased role of defects and peculiarities of a phonon spectrum. At $T \approx 1$ to 10 K, practically all metal-like nanomaterials demonstrate a high residual electric resistance and a low value of electric resistance temperature coefficient (ERTC). A significant change of the electric resistance appears, when $L \leq 100$ nm. Evaluations demonstrate that the specific electric resistance at the grain interface is $\rho_{ib} \sim 3 \times 10^{-12}$ Ohm cm and is practically the same for materials with the nano-and the big-size crystallite [1]. In such a way, the nanomaterial electric resistance may be calculated according to the formula:

$$\rho_\Sigma = \rho_0 + \rho_{ib}(S/V), \tag{6}$$

where: ρ_0 is an electric resistance of a single-crystalline material with a given content of impurities and defects; S is an area of the grain interface; V is a volume.

In addition, a porosity, an impurity content, and other factors should be taken into account in a determination of the electric resistance.

The electric resistance of a thin film depends on an electron scattering at an external surface, a topography, and structure features. An important role is played by the film thickness and the structure element size normalized to a free-range length.

Studies of a nanomaterial superconductivity, for example, of a refractory compound NbN, VN, TiN, NbCN, demonstrates a significant effect of the particle size on a critical magnetic field [17], and a decrease of temperature found in the course of transition to a super conductivity.

As it was noted above, a decrease of the grain size of the semiconductor resulted in an increase of the forbidden zone width to a level of a dielectric material (for example GaAs). Many other factors (like an origin, an increased number of segregations at the interface surface, a change in stoichiometry deviations, a perfect grain interface, *etc.*) also affect semiconductor properties. Therefore, a dependence of the electric resistance and the dielectric penetrability on the structure element size seems to be ambiguous.

Properties of a hybrid material, which is based on the nanocomposite, also attract interest. For example, a conductivity of a non-conducting matrix with metallic nanoparticles sharply increases when a certain percentage of conducting components is reached. It seems to be either due to barrier transition or to dominating tunneling (jump transition). A thermoelectric nanomaterial is characterized by an increased quality-factor (Q-factor). For such devices as a nanodiode, a nanotransistor, a nanoswitcher, *etc.*, an account of an increasing effect of the quantum factor (an oscillation, *etc.*) on a nanostructure conductivity is specially important.

1.5. MAGNETIC PROPERTIES

An actively studied magnetic characteristics are also sensitive to the structure element critical size. A keen interest is paid to magnetic properties of a small particle of the well known ferromagnetic material: an iron, a nickel, and a cobalt. A question arises: whether such properties as a coercitive field H_c, a magnetic anisotropy K, an initial magnetic permeability μ, a Curie temperature T_c, and others stay unchangeable, if a volume of the ferromagnetic material decreases to a negligibly low value of 10^3 to 10^4 atoms. What is a minimum amount of ferromagnetic atoms (for example, of the iron or cobalt), which should be picked-up together, in order that a particle would acquire ferromagnetic properties?

Before starting to discuss ferromagnetic properties of the small particle, let us consider a physical sense of applied terms-the Curie temperature and the magnetic anisotropy. A physical sense of the coercitive field H_c and magnetic permeability μ is reported in [18].

As it is known, a ferromagnetic state arises in a metal and an alloy due to an action of an electrostatic force, which orders a magnetic momentum of atoms. A domain, which is a region with a parallel magnetic atomic moment, is formed. When a ferromagnetic material is heated, thermal atomic motion gradually initiats its influence on the ordered arrangement of atomic magnetic moments, and fully destroys this arrangement at a certain temperature. A temperature, at which the oriented ordering of magnetic moments inside the domain is fully destroyed, is called the Curie temperature in an honor of an outstanding French physicist Pierre Curie. Above T_c, the ordered arrangement of magnetic moments is absent, and the ferromagnetic state is changed for the paramagnetic one.

Let us consider a concept of a magnetic anisotropy. Let us imagine such singlecrystal of the ferromagnetic material as an iron. Magnetic properties of this single crystal are strongly dependent on a direction of an applied external magnetic field due to the ordered atomic arrangement. Experiments demonstrate that two directions (in relation to magnetic properties) for the iron single crystal can be distinguished: when the crystal is easily magnetized along one direction (an axis of easy magnetization) and hardly magnetized along another (an axis of hard magnetization). This phenomenon is called a magnetic anisotropy. An anisotropy constant K is a difference in energies, which are spent to magnetize a unit of a ferromagnetic volume along axes of the easy and the hard magnetization. This K value for an iron is 4.2 x 10^4 J/m^3 at room temperature.

When a volume of a magnetic material decreases, the magnetic order undergoes a significant change. Conceptually, this may be understood as an increasing indeterminacy of mechanical moment ρ and electron energy occurring at a certain volume region d. Let us remember, that an electron is a carrier of a spin and orbital magnetic moment. A region of existing ferromagnetic state decreases according to a Heisenberg's relation $\Delta\rho = h/d$, where h is a Plank constant. When d is low, the energy is undetermined and, as a result, a long-range magnetic order is smoothed. The magnetic order of the iron is destroyed when $d = 1$ nm.

Experimental studies of ferromagnetic properties of a small iron, nickel, and cobalt particle demonstrate that its transition to a paramagnetic state dependes on a its size and temperature. The ferromagnetism disappeares when the iron or nickel particle size is about 6 to 7 nm. When the particle size is below these values, the material becomes paramagnetic. The iron particle of 6nm transferres to the ferromagnetic state only at $T_c = 170$ K. For a comparison, the Curie temperature of a bulky iron sample is $T_c = 1090$ K.

In such a way, experimental data demonstrate that the particle diameter "destroying" the long-range magnetic order essentially exceeds evaluations derived in the Heisenberg's relation ($d = 1$ nm for the iron). It is worth noting, however, that the above evaluations come closer to each other, if to take into account that an essential quantity of atoms occurs in a surface when the particle size decreases. Therefore, when the particle diameter is 2.5 nm, the quantity of atoms lieing in the surface exceeds 50% (it is assumed that the surface is covered by two layers of atoms). The nanoparticle coercitive field H_c also depends on the size. Fig. **8** [8] shows that 4nm crystallite has almost zero H_c value.

The low value of the coercitive field, which is conditioned by a heat effect, transforms the magnetic order into paramagnetic. A maximum H_c value of the nanocrystalline ferromagnetic material is observed when a particle is single-domain. Experimental data and theoretical evaluations almost coincide and demonstrate that H_c value for the iron with 20 to 25 nm crystallite size is maximal at room temperature.

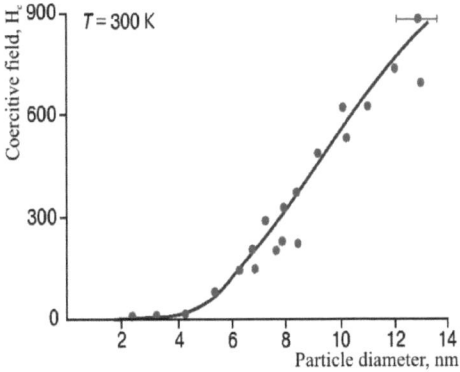

Figure 8: A change of a coercitive field *vs.* a particle diameter.

It follows from here, that the nanocrystalline ferromagnetic material is promising for a memory device of a high information density. For the first time in 1988, it was demonstrated that an iron-silicon alloy with the nanometer grain size and a random mutual orientation had a zero magnetic anisotropy. As it is known, that an absence of an anisotropy results in an essential growth of an initial magnetic penetrability both of a low and a high frequency. A little later (in 1991), $Fe_{73.5}Si_{13.5}B_9Nb_3Cu_1$ alloy with 10 to 20 nm grain size was fabricated [3] (figures indicate an atomic percentage of every element in this alloy). The alloy has also a low coercitive field ($H_c = 0.58$ SA/m), a high initial penetrability ($\mu = 105$), and a saturation induction $Bs = 1.25$ T.

A model presented in [10] explains a low magnetic anisotropy of a nanocrystalline alloy. According to this model, a macroscopic region of L area are formed in the ferromagnetic material. Magnetization inside these regions is such as if it follows a local magnetic anisotropy characterizing every crystallite of d-size. Taking into account a randomly directed magnetic anisotropy of every grain (a microscopic L size essentially exceeds d), an effect of the local anisotropy can be averaged, i.e. $\langle K \rangle \to 0$. When $L \ll d$, the resulting averaged anisotropy effect for an ensemble of randomly oriented and ferromagnetically interacting regions with d-size grains is $\langle K \rangle \sim d^{6}$. As it follows from this expression, this averaging could hardly allow a total elimination of the local anisotropy effect, but averaging at low d value essentially reduces its action. In this case, a role of this characteristic value (d) is very high, since d is introduced in this expression for the sixth order mean anisotropy. Experiments show that 10 to 20 nm size efficiently averages out the magnetic anisotropy effect. A decrease of the ferromagnetic anisotropy is accompanied by an increase of its magnetic penetrability and a decrease of its coercive field. Regularities of a change of an electromagnetic susceptibility occurring in a diamagnetic and paramagnetic material depending on the crystallite size are now not yet clear. However, a material having a high concentration of deformation defects demonstrates the increased magnetic susceptibility.

Magnetic properties essentially depend on an interatomic distance, therefore, it is natural to assume that a saturation magnetization I_s, a Curie temperature T_c, and other parameters characterizing the nanomaterial ferromagnetic and a big-sized crystalline object are be different. So, I_s value for the nanocrystalline Fe ($L =$ 6 nm) is lower by 40%. A superparamagnetism of nanoparticles (nanocrystallites) of a ferromagnetical, a ferrimagnetical, and an antiferromagnetical material becomes evident when an energy of heat fluctuations becomes comparable to a turn energy of a particle magnetic momentum. Table **4** schematically shows characteristics of various ferromagnetic materials, which are changed by an action of a corresponding field taking into account a dispersion.

The coercitive force H_c for various soft magnetic materials depends non-monotonously on the crystallite size. It increases and has a flat maximum up to 40 to 70nm. Then, it decreases together with the crystallite size. In general, optimal characteristics (a *min* H_c, a *max* magnetic penetrability $\mu = B/H$, and a saturation inductivity B_s) of a soft magnetic material are realized when the crystallite size is less than 20 nm.

Table 4: An effect of a size factor on characteristics of a ferromagnetic, a ferroelectric, and a ferroelastic material.

Crystallites size, nm	Ferromagnetics	Ferroelectrics	Ferroelastics
1000	Many-domain structure	Many-domain structure	Many-domain structure
100	One-domain structure	One-domain structure	One-domain structure
10	Super paramagnetics	Super paraelectrics	Super paraelastics
1	Paramagnetics	Paraelectrics	Paraelastics

A multi-layered film (a superlattice), which is composed of alternating nanolayers of the ferromagnetic and non-magnetic material (Fe/Cr, Co/Cu, Ni/Ag, *etc.*) and a powder of the nanocomposite material with the same composition, feature a giant magnetic resistance. When the magnetic field is applied, the electric resistance of these structures significantly decreases in comparison with a similar uniform material. A magnetic superlattice and a hard-magnetic material demonstrate transition to a superparamagnetic state when a film thickness and a crystallite size decrease. This transition is accompanied by a destruction of the magnetic order (a decrease of the magnetic characteristics).

A dependence of the magnetization on an applied magnetic field for the antiferromagnetic nanomaterial CuO and NiO has a typical form demonstrating hysteretic properties.

In addition, the magnetic material features a magnetocaloric effect *i.e.* an ordered magnetic spin of doped magnetic particles, which are introduced into a non-magnetic structure or a weak magnetic matrix along a direction of an applied magnetic field.

An increase of hardness and strength, a change of plasticity, elastic characteristics, stability, catalysis, and diffusion properties of the nanomaterial are now intensively studied. These properties may also be employed for a development of new elements of devices and open a possibility of transition to a new generation of materials, which properties can be radically changed by the grain size and the form of their structure elements.

QUESTIONS FOR CONTROL

1. What are nanotechnologies?
2. What is a place of nanoworld objects in a classification of general size scale?
3. What space sizes have nanoobjects?
4. What is a nanoconsolidated material?
5. To what an enhanced strength of the nanocrystalline material is related?
6. Describe a specific structure of a grain interface of a nanocrystalline material.
7. What is a fraction of nanocrystalline matter at a grain interface?
8. Cite the formulae describing a dependence of summary fraction of an interface surface, a grain interface, and a triple junction on a crystal size.
9. What are thermodynamic features of a nanostructure?
10. How is a nanomaterial electric resistance calculated?
11. What are features of a nanoferromagnetic material?
12. How does a ferromagnetism change under transition to a nanometer range?
13. What is a superparamagnetism?
14. What is a dependence of a coercitive force of a nanoferromagnetic material on a particle size?

REFERENCES

[1] Gleiter H. Nanostructured Materials: Basic Concepts and microstructure Acta Materialia 2002; 48: 1-29.
[2] Golovin YuI. Introduction in nanotechnology. Moscow: Mashinostroenie 2003.

[3] Andrievskii RA, Glezer AM. Grain Size effects in Nanocrystalline Materials. 1. Features of Structure. Thermal Dynamics. Phase Equilibrium. Kinetic Phenomena. The Physics of Metals and Metallography 1999; 88: 50-73.

[4] Andrievskii RA, Ragulia AV. Nanostructured Materials. Text Book. Moscow: Academiia 2005.

[5] Palumbo G, Erb U, Aust K. Triple Line Disclination Effect on the Mechanical Behavior of Materials Scripta Metallurgica et Materialia 1990; 24: 1347-1350.

[6] Gleiter H. Nanostructured materials: State of the Art and Perspectives. Zeitschrift fur Metallkunde 1995; 86: 78-83.

[7] Valiev RZ, Aleksandrov IV. Nanostructured Materials Fabricated by Intensive Plastic Deformation. Moscow: Logos 2000.

[8] Zhou Y, Erb U, Aust KT, Palumbo G. The Effects of Triple Junctions and Grain Boundaries on Hardness and Youngs Modulus in Nanostructured Ti-P Scripta Mater 2003; 48: 825-838.

[9] Gleiter H, Weissmuller J, Wollersheim O, et al. Nanocrystalline Materials: A Way to Solids with Tunable Electronic Structure and Properties. Acta Materialia 2001; 48: 737-745.

[10] Zhang S, Sun D, Fu Y, Du H. Recent Advances of Superhard Nanocomposite Coatings: A Review. Surf Coat Technol 2003; 167: 113-119.

[11] Liakishev NP, Alymov MI. Nanomaterials for Structured Materials. Nanotechnologies in Russia 2006; 1: 71-81.

[12] Ragulia AV, Skorohkod VV. Consolidated Nanostructured Materials. Kiev: Naukova Dumka 2007.

[13] Andrievskii RA. Nanomaterials: A Concept and Modern Problems. Rus Chem Jour 2002; 46: 50-56.

[14] Andrievskii RA, Glezer AM. Grain Size Effects in Nanocrystalline Materials. II. Mechanical and Physical Properties. The Physics of Metals and Metallography 1999; 88: 50-73.

[15] McGrea JI, Aust KT, Palumbo G, et al. Electrical Resistivity as Characterization Tool for Nanocrystalline Metal In: Kommarneni EdsS, Parker JC, Hahn H. Nanophase and Nanocomposite Materials III. Warrendale Materials research Society 2000; 461-466.

[16] Troitskii VN, Domashnev IA, Kurkin EN, et al. Plasmo-Chemical Synthesis and Properties of Ultradispersion NbN. High Energy Chemistry1994; 28: 275-279.

[17] Zolotukhin IV, Kalinin YuV, Sognii OV. New Directions of Physical Material Science. Voronezh: VGU 2000.

[18] Newnham R. Size effect and NonLinear Phenomenon in Ferroic Ceramics. Third Euro-Ceramics eds P.Duran, J. Fernandes Faenza Editrice Iberica 1993; 2: 1-9.

[19] Kolobov YuR, Valiev RZ, et al. Diffusion at Grain Interfaces and Properties of Nanostructured Materials. Novosibirsk: Nauka 2001.

[20] Nalwa HS. Nanostructured Materials and Nanotechnology St.-Peterburg: Nauka 2001.

[21] Reithmaier JP, Kulisch W, Petkov P. Nanostructured Materials for Advanced Technological Applications. Netherlands: Springer 2000.

[22] Knauth Ph, Schoonman J. Nanostructured materials. Selected Synthesis Methods, Properties and Applications New York: Kluwer Academic Publishers 2004.

[23] Peidong Y. Chemistry of Nanostructured Materials. World Scientific Publishing Company 2004

[24] Advani GS. Processing and properties of nanocomposites. World Scientific Publishing 2007

[25] Aliofkhazraei M. Nanocoatings: Size Effect in Nanostructured Films. Springer 2011

[26] Bhushan B. Handbook of Nanotechnology. Springer 2010.

[27] Rodriguez JA, Fernandez-Garcia M. Synthesis, Properties and Applications of Oxide Nanomaterials. Wiley 2007.

[28] Kanellopoulos N. Nanoporous Materials: Advanced Techniques for Characterization, Modeling, and Processing CRC 2011.

[29] Books New Nanotechnologies/edit. A. Malik and R.J. Rawat. A.D.Pogrebnjak, A.P.Shpak, V.M.Beresnev. Chapter 2 (p. 25-114). Structure and Properties of Protective Composite Coatings and Modified Surface Prior and After Plasma High Energy Jets, *Nova Science Publisher*. 2009; 687.

[30] Pogrebnjak AD, Lozovan AA, Kirik GV, et al. Structure and Properties of nanocomposite, hybrid and polymers coatings,Publ. House URSS, Moscow, 2011, 344.

[31] Azarenkov NA, Beresnev VM, Pogrebnjak AD, et al. Fundamentals of Fabricated Nanostructured Coatings, Nanomaterials and Their Properties, Publ. House URSS, Moscow 2012; 352.

CHAPTER 2

Nanoporous Materials

Abstract: This Chapter describes a whole complex of nanoporous materials. A classification of porous material (an inside pore, an opened pore, a perforating pore, an opened-line-end pore) is considered. Three different ways of interaction of the nanoporous material with a surrounding medium are demonstrated: a-adsorption; b-filtration, c-catalysis using: 1) small molecule and 2) big molecule. The problem of how to control size, form of pores, and uniformity of their distribution is considered.

Keywords: Types of nanoporous materials, applications.

2.1. INTRODUCTION

Formally, a nanoporous material may be considered as a nanocomposite, in which a pore plays a role of a second phase, which is randomly or regularly distributed in a matrix (Fig. **1**). However, there are several physical reasons to single out the nanoporous material as an individual class of materials.

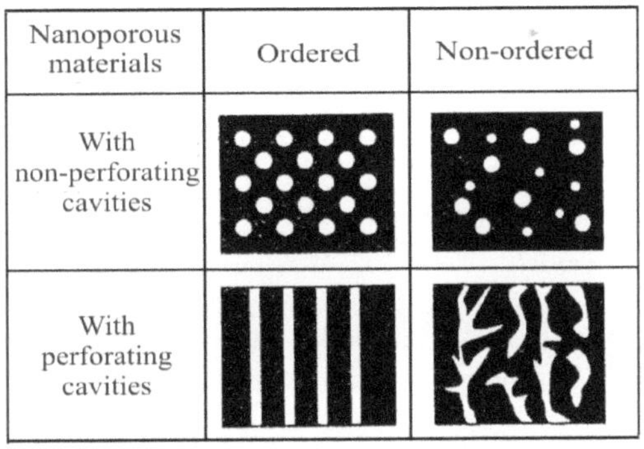

Nanoporous materials	Ordered	Non-ordered
With non-perforating cavities		
With perforating cavities		

Figure 1: Main types of a nanoporous material.

A detailed consideration of genesis, behavior, and influence of pores on solid properties seems to be more reasonable to start from determination of free volume and porosity concepts, since every of these two concepts is not always unambiguous [1-15] if to take into account multiple variations of existing nanoporous systems, when a universal criterium of their description is absent.

A free volume. Let us define a free volume V_f in a solid as a space, which is not filled with atoms, and an electron density of which is close to zero. A distribution character of the free volume is conditioned by a degree of atomic packing uniformity over this volume and may be set in the course of analysis of an electron density distribution function. A concretization of the free volume concept, taking into account a way of its determination, is important. Therefore, a geometrically free volume V_r is

$$V_r = V \Sigma \omega, \qquad (1)$$

where $\Sigma\omega$ is a summary specific atomic value, which characterizes a solid volume, which was occupied by a "void" phase. In this case, a pore may be considered as a local free volume unit of a definite form, location, sizes of which essentially exceed interatomic distances of a material matrix. Then, a relation of solid summary volume $V-V_k$, which was occupied by the "void" phase, to a full volume characterizes a volume fraction (a percent volume concentration) of pores, or a solid porosity, C. A value, which is opposite to the porosity, is called a degree of occupation (population) density. A volume of compact solid of chemically identical composition, a mass of which is equal to a porous solid, is designated as V_k.

If $\Sigma\omega/V$ is an atomic packing coefficient, then a relative geometrical free volume represents a value $1-K_{pack}$.

Also, various types of interstitial voids, which, however, according to selected criteria, cannot be considered as pores, are integral components of the free crystalline volume.

To identify a porosity term, two approaches are applied. One of them is based on an idea that pores are characteristic and integral components of structure and define an origin, properties, and purposes of material. This approach is usual for objects containing essential number of voids (soils, foam materials, fabrics, activated coals, gels, zeolites, porous ceramics, certain kinds of fine-fibred structures, porous sintered composition materials). In this case, they speak about a porous state of material, since an individually considered pore cannot be clearly defined with respect to some of the above mentioned systems. Therefore, a statistical geometry, a stereology, methods of statistical and stochastic physics, methods of molecular analogy, as well as empirical methods with a theory of random functions, *etc.* are applied to describe them. To facilitate studies of morphological characteristics, which specify a course of various processes in a porous medium, real porous systems are classified according to *a formation mechanism* (growing systems or substance porous systems, addition(composition) systems, as well as combined systems) and a *structure character* (systems with an ordered and non-ordered arrangement of structure elements).

A growing system is formed when a macroscopically continuous medium is dispersed in the course of sublimation, condensation, and solidification and/or biological processes. Solidified foams, sponges, cokes, activated coals, pumices, zeolites, fine condensed layers of thermally stable materials, which were deposited to chemically neutral substrates under a significant overcooling and an absence of chemical adsorption, fibers of cellulose, skeletons of certain plants and organisms are among them.

An addition system is formed when a reasonable number of elements, a specific porosity of which may be neglected (for example, a colloid system (a gel in a dry state), a granular material, a filter fiber, a yarn, a paper, *etc.*) are added occasionally and randomly. A combined (or complicated) system is a combination of growing and addition system, in particular, a product of pressing and sintering applied in a powder metallurgy, a ceramics, a fabric, a building material, a porous glass, and others, in which a formation of pores formation is preceded by a random or regular addition (an elimination) of individual elements of the given system. Various condensates, which are deposited under conditions of overcooling, can also be attributed to such systems, if they are unfairly frozen and there is a possibility of structure transformations during a post-condensation period.

A consideration of pores as three-dimensional imperfections (defects) of solid structure together with zero-, one-, and two-dimensional defects is one more approach, which is applied for a determination of porosity. Such presentation of pores works for materials, which origin and purposes are not directly related to pores as characteristic structure "components". Cast metals and alloys of low porosity, rolled stock materials, the majority of minerals and glasses, *etc.* can be attributed to such materials.

However, one should bear in mind that terms "a low porosity" and "a high porosity" are conventional, and their correctness is evident only in the case of statistic methods taking into account a dispersion ability, a form, an orientation, and a dimensional distribution of pores. Analyzing an effect of pores on material properties, one does not need this conventional division, because a sensitivity of properties to pores, their number, a dispersion, a form, and a volume distribution character is different. Both above-mentioned approaches may be changed for a single one when pores are considered as foreign phase inclusions or phase-structure non-uniformities in solids. This is convenient for a thermodynamic description of porous systems and their diagram presentation. This approach allows an evaluation of lability of any system and determination of optimum thermodynamic parameters for states, when a given system is truly considered as a conventionally "solidified" one.

Morphological Characteristics of Pores. Morphological features of pores as volume non-uniformities are determined by their construction (Fig. **2**).

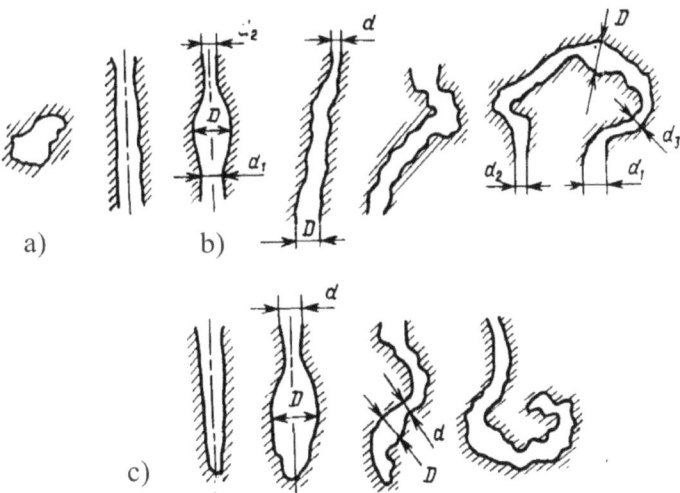

Figure 2: Types of pores in solids: *a*-an inside (closed) pore; *b*-an opened throughout pore; *c*-an opened dead end pore.

An isolated body closed in a volume, a solitary non-continuity, and a "sealed" pore ensemble, which may have a mutual communication, refer to an inside pore. A surface of an inside pore is not taken into account in calculations of overall body surface. An open pore is a throughout *(b)* and a dead-end *(c)* pore representing capillaries of arbitrary forms closed from one side. In such a way, a body overall porosity *C* is a sum:

$$C = C_{in} + C_{op} = C_{in} + C_{th} + C_{d}, \qquad (2)$$

where C_{in} and C_{op} is an inside and an open porosity; C_{th} and C_d is a volume fraction of throughout and dead-end pores. Pores, which in a general case have arbitrary forms and dimensions, may be localized both inside solid structure elements (for example, inside crystallites, fragments, blocks, cells, or granules) and along their boundaries, depending on a matter prehistory, an energy balance, and a structure. A chaotic and an ordered arrangement of pores is distinguished by an analogy with a modulated structure, which is related to a formation of pore superlattices (simple and complicated) or subsystems with periodically changing hollow cells of various ordering degree. In many materials (for example, anisotropic deposited films, composites, and other materials, which were subjected to an oriented action) a pronounced dominating orientation of pore arrangement can be observed.

A great number of small pores and channels (their transverse sizes may vary from 0.3-0.4nm to a micrometer range) imparts a number of special physical properties to nanoporous materials. Methodically, according to a classification of International Chemical Union, porous materials are divided into three classes: microporous (with a characteristic pore size $R < 2$ nm), mezzoporous ($2 < R \ll 50$ nm), and macroporous ($R > 50$ nm) materials. This differs, a little, from a classification adopted for nanostructural materials in a material science, because a difference between these mentioned groups is shifted towards lower R. However, the division itself, its principles, and consequences remain the same. A free surface accessible for an interaction with gases and liquids exceeds a surface of continuous solid by orders of magnitude reaching 1000m²/g. This leads to an enhancement of conditions for a heterophase chemical and catalytic reaction, amends a sorption capacity, *etc*. However, a simple increase of specific surface is not the only reason of increased activity characterizing nanoporous materials.

A relatively high amount of atoms arranged in a surface itself and high curvature near-surface layers may drastically change properties of material itself, properties of molecules and atoms, which were adsorbed from a surrounding medium by its pores.

An important characteristic of porous bodies is also their penetrability for gases and liquids. In the case of a nanorange transversal pore size, it may vary adapting itself to molecules of different forms and sizes, *i.e.* a nanoporous material may be used as a selective molecular sieve and filter (K is a catalyst particle; A and B are initial reagents; $A + B$ is a synthesized product) (Fig. 3).

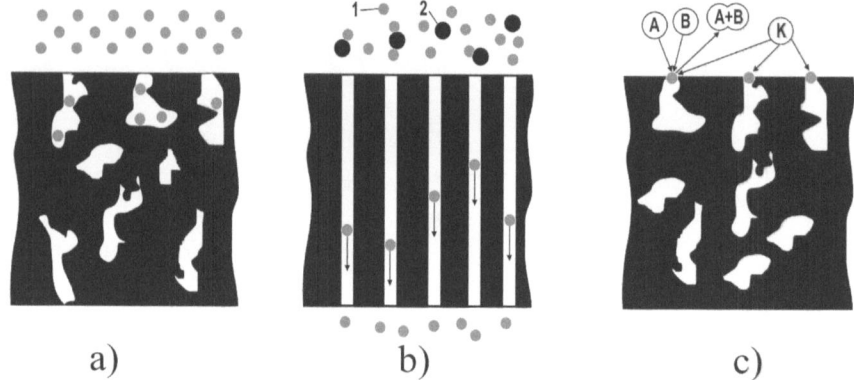

Figure 3: Three main types of interaction of nanoporous materials with a surrounding medium: *a*-an adsorption; *b*-a filtration, a separation of mixtures; *c*-a catalysis; *1*-small molecules; *2*-big molecules.

Porous penetrable materials are widely employed as filters for a mechanical purification, a drying, a heating of liquids and gases, a selected filtration, a separation, and an enrichment of gases, an aeration of liquids and powder materials, reactions in a boiling layer, a dosing and a uniform distribution of liquids or gases, an evaporization cooling of constructions working under high-temperatures, mixture regenerators, dust collectors, humidifiers, desalters, nuclear membrane filters, *etc.*

A character and a kinetics of these processes for gases and liquids flowing through a porous media are determined mainly by a size ratio of pores and fluid molecules, an action of adsorption and electrostatic forces, a concentration, a form, and an orientation of pores, their dimension distribution over material, as well as an interphase tension at a nanopore output.

A rate, with which a liquid is sucked into the nanopore of R radius, can be approximately evaluated from the following expression:

$$l_{liq}/\tau = R^2 \rho_{liq}/8\eta_{liq} \{(2\alpha\cos\theta/R\rho_{liq} R) - g\sin\beta_{edg}\}, \tag{3}$$

where l_{liq}-is a length of liquid column in the nanopore, which was absorbed for τ time, η_{liq} and ρ_{liq}-are a viscosity coefficient and a specific weight of absorbed liquid; α and θ-are a liquid surface tension and an edge angle of wetting; β_{edg}-is a slope of nanopore axis to a horizon; g-is a gravitation acceleration. However, according to the formula *(3)*, the calculation of sucking-in rate for a complicated system of branching nanopore is only a reference point.

Main qualities of filtering porous materials are their permeability, a degree of fine cleaning, a retaining ability, and a mechanical strength, a chemical and a thermal resistance for filtration of aggressive fluids and an operation under enhanced temperatures, as well as their chemical and thermal resistance. The degree of material permeability is determined by a fluid velocity flowing through a unit of area and a unit of material thickness under a given pressure. Three types of fluid transportation through a porous medium are distinguished. *A diffusion* is induced by an action of concentration and temperature gradients under a constant pressure. *An effusion* is a free-molecular flow (Knudsen Flow), when an impact frequency of fluid molecules is negligible, in comparison to an impact frequency to the nanopore surface. *A viscous laminar flow* or *a transpiration* follows a convection mechanism, when a fluid is flowing like a continuous medium under an action of pressure gradient and intermolecular impacts dominate over impacts of free surface molecules.

From the point of view of a basic material or a technology of nanoporous materialfabrication, they may be divided into a ceramics, a semiconductor, a polymer, and a biological material. They are applied for gas masks, autonomous life support systems of submarines and habitable space ships, fine cleaning filters for an air and water (including a filtration of chemical and bacteriological toxic agents). For their efficient action, it is important to overcome a *percolation limit*, out of which pores and channels having no earlier communications begin to form through-passes for a fluid medium. Materials having through-passing channels of regular geometrical forms and similar sizes have a especial value for these applications, since the lower is their difference, the better are filter properties.

In a chemical, metallurgical, and biotechnological industry, one of the most popular types of nanoporous materials is a *ceolite*-an alumosilicate, which is fabricated from a special clay. Pores of about 0.1 to 10 nm size, with a three-dimension structure and through-passing channels are formed after a special thermal treatment. A pore size depends on a number of oxygen atoms in a cyclic structure, which form the ceolite allowing an easy absorption of definite molecules or a separation of their blends in a membrane filter.

Last century, at the beginning of 90th, Mobile Oil OH firm reported about a finished development of a new class of alumocilicates (CMC-41, *etc.*) containing ordered cylindrical pores of 2 to 10 nm diameter with a low size difference and a high separation ability.

A key problem in a fabrication of porous nanomaterials is a control of pore size, its form, and spatial distribution uniformity. The Institute for Material Science of NAN of Ukraine offered a simple method for a fabrication of porous materials with a pronounced submicron microstructure. The method is based on a self-construction of nucleus-shell structures using a heterophase coagulation. A monodispersion submicron polymer sphere was applied as a nucleus template, and a nanosized ceramic particle-as *material model*. A heterocoagulation made possible a nanoparticle surface modification and a polymer mixing. A nucleus-shell structure started its self-construction under an action of Coulomb forces. Packing laminated particles formed a closely packed structure under a vacuum filtration,. Then the polymer was disposed in the process of calcinations, as a result of which they formed a desired porous structure. The pore size may be controlled by changing a size and a form of polymer *nucleus*. A porosity can be easily specified by a volume ratio of polymer/ceramics particles. This process allows a formation of various porous structures.

An application of porous materials is highly important for a transformation of exhaust gases, a purification (a refinement) of liquids, a catalytic reforming employed in a chemical industry, *etc.* Properties of porous materials properties can be optimized by a control of pore structure. So, for example, micropores should have an equal size when a porous aluminum oxide Al_2O_3 is used as a carrier of silver catalyst (which is used in an ethylene fabrication). A microfiltration on ceramic membranes is a promising application of nanoporous aluminum oxide. However, a typical ceramic membrane-fabricated by a sintering of nanodispersion aluminum oxide powder demonstrates a relatively low permeability due to a relatively low porosity. To fabricate the membrane filters of high permeability, we needs some alternative method, which is a sol-gel condensation or a deposition from chemical vapors. The deposition of ultradispersion powder to a porous substrate also is an alternative for the membrane filter.

A great interest was attracted by a discovery of the 90th-a luminescence of porous silicon in a visible light. It was absent in an ordinary silicon singlecrystal, which, as it was known, detected only a weak fluorescence of 1.0 to 1.2 eV quantum energies in an infrared spectrum with about a forbidden zone width. In addition to the photoluminescence, the porous silicon demonstrated also an electroluminescence, a cathode luminescence (*i.e.* a visible light emission induced by an applied voltage and by electrons falling to its surface).

Porous materials attract a great interest, since they open a possibility to combine nanoparticles, which are arranged in mezzopores.

QUESTIONS FOR CONTROL

1. What are the types of nanoporous materials?
2. What characterizes a porosity?
3. What is a key problem now in a fabricationof nanoporous materials?

4. What types of interaction exist between nanoporous materials and a surrounding medium?

5. What is a ceolite and where is it applied?

REFERENCES

[1] Cheremskoi PG, Slezov VV, Betekhin VI. Pores in Solids. Moscow: Energoatomizdat 1990.
[2] Shpak AP, Cheremskoi PG, Kunitskii YuA, Sobol OV. Cluster and Nanostructured Materials. Kiev: VD Akademperiodika 2005; 3: 516.
[3] Denisova NE, Shorin VA, Gontar IN, *et al.* Tribological Material Science and Tribology. Penza: PGU 2006.
[4] Ragulia AV, Skorokhod VV. Consolidated Nanostructured Materials. Kiev: Naukova Dumka 2007.
[5] Sokolov S, Bell D, Stein A. Preparation and Characterization of Macroporous Alpha-Aluminum. J Amer Ceram Soc 2003; 86: 1481-1486.
[6] Kanellopoulos N. Nanoporous Materials: Advanced Techniques for Characterization, Modeling, and Processing CRC 2011.
[7] Roque-Malherbe Rolando MA. Adsorption and diffusion in nanoporous materials. CRC Press Taylor & Francis Group 2007.
[8] Sattler KD. Handbook of Nanophysics: Nanoelectronics and Nanophotonics. CRC Press 2010.
[9] Shenderova OA, Gruen DM. Ultrananocrystalline Diamond: Synthesis, Properties, and Applications. NY: William Andrew Publishing 2006.
[10] Steiner T. Semiconductor Nanostructures for Optoelectronic Applications. Artech House Inc. 2004.
[11] Yangyang S. Study on the nanocomposite underfill for flip-chip application. Biochemistry Georgia Institute of Technology 2006.
[12] Sung J, Lin J. Diamond Nanotechnology. Syntheses and Applications. Pan Stanford Publishing Pte. Ltd. 2010.
[13] Vo-Dinh T. Nanotechnology in Biology and Medicine. Methods, Devices, and Applications. CRC Press, Taylor & Francis Group 2007.
[14] Pogrebnjak AD., Uglov VV, Il'yashenko MV, *et al.* Nano-Microcomposite and Combined Coatings on Ti-Si-N/WC-Co-Cr/Steel and Ti-Si-N/(Cr3C2)75-(NiCr)25 Base: Their Structure and Properties Nanostructured Materials and Nanotechnology IV: Ceramic Engineering and Science Proceedings 2011; 31(7):115-126.
[15] Pogrebnjak AD, Sobol OV, Beresnev VM, *et al.* Phase Composition, Thermal Stability, Physical and Mechanical Properties of Superhard On Base Zr-Ti-Si-N Nanocomposite Coatings Nanostructured Materials and Nanotechnology IV: Ceramic Engineering and Science Proceedings 2010; 31(7): 127-138.

CHAPTER 3

Amorphous Materials

Abstract: A definition of amorphous material is presented. It is considered as a transition from an amorphous to a nanocrystalline state. An atomic structure of amorphous body is considered. Mechanisms and classes of amorphous bodies, which are genetically bound with crystals or liquids, are briefly described. Methods employed for an investigation of the amorphous structure are described. A atomic structure of glasses is considered from the point of view of radial distribution functions (RDF). A model of atomic structure free volume for the amorphous materials is presented. Properties of amorphous materials are briefly described.

Keywords: Polycluster, models properties of amorphous metallic.

3.1. INTRODUCTION

Recently, an amorphous solid attracted a great interest of scientists working in the field of fundamental and applied physics. An absence of long-range order in a mutual atomic arrangement is a determinant point of amorphous body. Such system is characterized, on one hand, by the absence of long-range order, *i.e.* a strict periodicity of atomic arrangement in a microvolume, on the other hand, by a presence of short-range order, *i.e.* an ordered distribution of coordinates of nearest neighbors for any atom. The absence of long-range translation order often leads to an alternation of properties, which are difficult or impossible to reach in a solid with a crystalline structure. Some of these properties turned out to be very important both for practical applications and for a scientific understanding of atomic disorder.

An application of amorphous solid attracts a great interest since it is a material with preliminarily "given" properties. First, an optical glass started to draw a great interest for fiber-optical communication systems. Then, an amorphous metallic alloy (a metallic glass) found its application as a magnetic cap core, a magneto-mechanical pickup, a regulated delay line, *etc.* An amorphous semiconductor is a photodetector, a relatively cheap solar battery, a sensitive layer for a xerography. The practical interest dictates a necessity to fabricate a new non-crystalline material. The following relation is necessary in order to reach desired macroproperties: *a composition-fabrication conditions-a structure-properties*. Today, scientists, who are working in the field of solid state physics and physical-chemical material science, are involved in this problem.

An amorphous state-is one of the existing forms of solids. The amorphous solid is a thermodynamically nonequilibrium or a metastable system tending to gain a crystalline structure and to go to a stable state. Regions with a definite atomic order (nucleuses) may arise in the metastable state due to fluctuations.

On one hand, if nucleus size exceeds certain critical value (critical radius), free system energy gets lower with growing nucleus. On the other hand, if nucleus size is lower than this critical value, free system formation and its growth need higher free energy in comparison with initial equilibrium state.

A time of metastable state ordering depends on a nucleation rate-a formation frequency of nucleuses exceeding a critical size, which is normalized to a volume unit and on a growth rate of supercritical nucleuses. An ordering time of non-equilibrium amorphous body depends on a local reconstruction frequency of atomic configurations resulting to the ordering.

An atomic structure of amorphous body depends not only on a character of interatomic forces, but also on the formation conditions. There are two classes of amorphous bodies, one of which is genetically related to crystals, another one-to liquids. If a crystal contains random dislocation nets or if it is a polycrystal composed of randomly oriented crystallites, then pair correlation radii of their lattice sites are comparable to an average distance between dislocations and crystallite sizes. Extended defects-dislocations and grain interfaces-play a dual role: first, they introduce a topological disorder destroying correlations in atomic arrangements at distances, which are comparable to average distances between defects, and, second, they destroy a local order of dislocation nuclei and boundary layers. In addition, dislocations are surrounded by fields of elastic deformations slowly decreasing with distances (like L/r, where r is a distance to a dislocation). A Bloch theorem is not applicable and electron wave functions (this is valid also for other quasiparticles) cannot be described by Bloch functions for a topologically disordered crystal. A good accuracy description of disordered crystal electron properties can be obtained if a topological disorder is neglected, the Bloch functions are taken as initial wave electron functions in a zero approximation, and a scattering at defects is taken into account. In such a way, a satisfactory description of considered amorphous bodies can be obtained if the topological disorder generated by extended defects is ignored and a resulting local disorder within the framework of perturbation theory is taken into account.

When a temperature exceeds a crystal melting point T_m, a liquid is a thermodynamically equilibrium condensed body. It may transfer to a metastable state if a temperature abruptly falls below T_m. First, the crystallization rate of overcooled liquid increases with a degree of overcooling ($\Delta T = T_m-T$) and, then, quickly decreases. So, a deep liquid overcooling transforms a liquid into an amorphous solid having an enormous crystallization time. Atomic configurations of this overcooled liquid are reconstructed both due to atomic diffusion displacements and changes of atomic interactions. Since atoms have no time for a serious reconstruction by a diffusion mechanism in the course of fast cooling, and average interatomic distances (which have a little difference both for a liquid and for a crystal) and atomic interactions do not essentially change, local topological changes of atomic configurations will not be great, and a structure of resulting solid and initial liquid will be similar. Many overcooled liquids have a characteristic temperature- a so-called glass-transition temperature T_g, when a viscosity is sharply increased, a specific heat capacity and a density decreased. Fig. 1 shows schematically a behavior of specific heat capacity and body viscosity for various aggregate states. The overcooled liquid when $T < T_g$ is called a glass, where T_g is considered as a temperature, at which the overcooled liquid is transformed into the amorphous solid. Fig. **1** shows a temperature dependence of specific heat capacity. Within (T_g, T_m) temperature interval the liquid is overcooled. Arrows indicate directions of temperature changes.

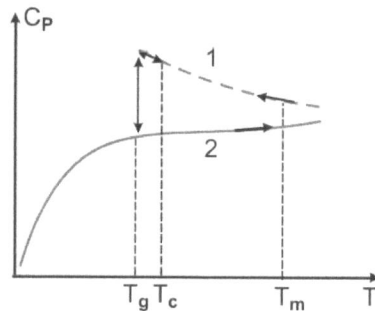

Figure 1: A dependence of specific heat capacity on a temperature: *1*-for a liquid and a glass; *2*-for a crystal.

However, as a rule, the glass-transition temperature depends on a cooling or heating rate, and, in most cases, changes at about T_g are not abrupt and are partly irreversible. A jump of C_p indicates that when $T \approx T_g$, the solid inside freedom degrees abruptly change. The irreversibility of these changes indicates a structure non-equilibrium state. Many glasses, first of all, metallic glasses (amorphous metals and alloys obtained by a fast quenching from a melt) are crystallized at T_c temperature, which a little exceeds T_g. In some cases, crystallization prevents the glass-to-liquid transition, because this transition temperature interval is smoothed or T_g coincides with a crystallization temperature.

In spite of an indefinite and a conventional character of a glass-transition temperature and a glass-to-liquid transition, this narrow temperature interval, which is characterized by essential changes of thermodynamical values and mechanical properties, allows an identification of matter state (liquid or solid).

Consequently, a cooling liquid has to reach such temperature level, at which a kinetic factor decreasing a formation rate of crystallization centers with a temperature will dominate. In a literature, terms "an amorphous state" and "a glass-like state" are often mentioned and understood as synonyms. Indeed, these terms are very close but "the amorphous state" is common. One can say that any glass-like state is amorphous, but not any amorphous state is a glass [1-4].

A main peculiarity of glass-like state, which differs it from other amorphous states, is that a glass can have a reversible transition from the glass-like state to a melt and from the melt to the glass-like state. This property is characteristic only for the glass. In the course of heating, other types of amorphous states, first, start a transition to a crystalline state and only when the temperature increases to a melting one, they start a transition to a liquid state. A gradually increasing viscosity of glass-forming melt prevents the crystallization, *i.e.* the transition to a more stable thermodynamic state with a lower free energy. Glasses are characterized by a number of physical properties. Practically, all glasses have a weak luminescence. Local mechanical stresses and structure non-uniformity yield a double refraction. As a rule, glass-like matters feature diamagnetic properties, and a doping of rare-earth metal oxides makes them paramagnetic materials. In contrast to a glass-like state, a metallic amorphous matter does not feature the reversible transition to the amorphous state (a melt). When the matter is heated, first, it transits to a crystalline state, and subsequently- to a liquid one. When it is cooled, in order to reach a crystalline state, the matter needs strictly definite conditions. The amorphous metallic state of matter does not demonstrate luminescent features. The majority of such matters is either a ferromagnetic or an antiferromagnetic material.

A structure of amorphous metallic system (AMS). The amorphous solid state is one of the most poorly studied fields of a modern structure material science. The main difficulty is a way employed to a description of amorphous state structure, since an absence of translation symmetry elements and concepts of elementary cells deprive a researcher of terms and concepts accepted in a crystallographic science, as well as powerful instruments for a structure analysis. The amorphous solid structure is similar to a liquid one, therefore, density fluctuations, local surroundings, and a chemical composition should be taken into account. This introduces a probabilistic and statistical character into the structure description.

An amorphous solid structure is studied using an X-ray, electron and neutron diffraction, and an *EXAFS* method (Extended X-Ray Absorption Fine Structure). Also, important methods are an electron microscopy, a small-angular scattering of neutrons, an X-ray analysis, measurements of Mossbauer effects, and a nuclear-magnetic resonance. The structure state and changes are reflected in magnetic, elastic, non-elastic, electrical, and other properties, which are studied using an *AMS* method [5-9, 17-23].

So for example, a structure of disordered system in a thermodynamic equilibrium state (a gas, a liquid) may be described with the help of distribution functions in a one-, two-or many-particle approximation. As for a theoretical description of non-equilibrium system structure, any reliable systematic approach is still unavailable. The *AMS* helps to determine a spatial atomic arrangement by an X-ray (or neutron) scattering intensity using an integral Fourier transformation. An information about the spatial atomic arrangement in a solid is taken from a structural model. Today, a direct observation of glass atomic structure is impossible. Indirectly, it can be determined using statistical characteristics such as a radial distribution function *(RDF)* and others. Experimental observations indicate a notable *RDF* similarity of a liquid and a glass, which confirm their structure similarity.

Thus for example, the radial distribution function *(RDF)* for a system composed of one sort of atoms is determined by the expression:

$$W(r) = 4\pi r^2 \rho(r), \tag{1}$$

where *r* is a distance from a picked-up atom; *ρ(r)* is an atomic distribution function or a twin atomic distribution function.

This function has the following physical sense. If a coordinate system is aligned with a center of one particle, a product *ρ(r)dV* is an average number of particles displaced in a volume unit *dV* with a characteristic radius-vector r. A function $W(r)$ is a weight number of atoms in a spherical cell of an r-radius and a layer thickness of unity.

When the r value is lower than the atomic sphere diameter, this function is equal to zero. And when the r distance increases, a correlation between particles is gradually dying and tends to an average value $\rho_0 = N/V$ when $r \to \infty \rho(r)$. Here, N is the number of particles in a volume V, The RDF is characterized by a first high peak corresponding to nearest neighbors, which, in their turn, correspond to the second, third, *etc.* neighboring particles around the picked-up atom. Most often, instead of $W(r)$ function, the above mentioned radial distribution function $G(r) = 4\pi t \{\rho(r)-\rho_0\}$ is used, which is a twin function of radial distribution $g(r) = \rho(r)/\rho_0$ and an interference function $J(k) \sim J_n F_2$, where f is an atomic factor and J_n is a scattering intensity.

In a description of amorphous solid containing atoms of different elements (n), the RDF is expressed as:

$$W(r) = 4\pi r^2 \Sigma \Sigma \omega_{ij} \rho_{ij}(r). \tag{2}$$

The expression includes a partial function of atomic distribution $\rho_{ij}(r)$ with a definite weight multiplier ω_{ij}. The partial function of atomic distribution $\rho_{ij}(r)(i, j = 1, 2, ...n)$ represents an average particle density of the ith sort in a volume unit dV of an r coordinate, if a particle of the ith sort occurs at a point of an $r = 0$ coordinate.

An area under the first RDF peak allows a determination of the average coordination number, *i.e.* the number of nearest neighbors in a single-component amorphous system:

$$Z = \int_0 \rho(r) 4\pi^2 r \rho d\rho \tag{3}$$

An average interatomic distance for the given structure can be derived from the first maximum position. Here, r_{min} is the first RDF minimum.

For a multi-component system, a calculation of short-range order parameter is complicated, since one needs a sum of atomic distribution partial functions with their weight multipliers, which indicate a relative contribution of individual components to a total interference function or the total RDF. A comparison of experimentally measured RDF for a liquid and an amorphous metal shows a satisfactory similarity.

Models for an amorphous solid. Today, there is a number of models describing a structural state of amorphous solid. Among them, a Bernal model (a *free space model* (FSM)) is mostly developed and widely employed to explain the amorphous solid properties. Every atom occupies a void among its nearest neighbors. If this void v_h exceeds a certain value v_c, which is approximately equal to a space of one atom in a closely-packed structure, then

$$Vr = v_h - v_c \tag{4}$$

This model was initially derived to explain a diffusion in a liquid. Later, it was developed and applied to describe diffusion properties and a plastic current of amorphous solid [9-16].

In the process of further development, the FSM was used for a description of plastic deformation and a solid state-liquid transition in an amorphous solid, as well as an explanation of tunneling state origin and a structure change kinetic occurring in the process of annealing.

Assumptions about cells containing a free space and premises about a density and a distribution of these cells under various conditions, as well as types of the relating structure reconstructions are significant elements of FSM. Here, details concerning configurations of "normal" cells and regions containing an excessive space are insignificant. This allows a formulation of free space model based on a representation that cells containing the free space play a crucial role in a plastic deformation, a diffusion, and a structure relaxation of glasses. This model has been developed in details and is now widely applied in a physics of amorphous solids. The free space model is based on the following assumptions:

- A free space may be redistributed between cells and this distribution occurs without a change of summary free space and a solid free energy;

- An atomic transfer by a diffusion is possible when a neighboring free space exceeds a certain critical value g^* comparable to v_c and this space was formed by a fluctuation-induced redistribution;

- An amount of free space may change when a temperature changes near T_g and as a result of stress action inducing a plastic deformation (T_g is a glass-transition temperature, at which a viscosity sharply increases but a specific heat capacity and a density decrease).

There are two interpretations of RDF. The first one comes to a consideration that an amorphous solid is considered as a system of close-packed stochastic spheres with an excessive (in comparison with a crystalline solid) specific space per every atom. This excessive atomic space or interatomic voids, may be randomly redistributed between atoms without a change of summary excessive space. A distribution

$$P(v_f) = \{\gamma/v_f\} \exp(-v_f/v_f) \tag{5}$$

seems to be the most probable.

Here, P is a distribution function; γ is a geometrical factor of unity order; v_f is a free space per one atom, and

$$v_f = \alpha (T-T_0), \tag{6}$$

where α is a heat expansion coefficient and T_0 is a phenomenological parameter.

Taking into account that an atom may be displaced to a void, a volume of which exceeds v^*, a diffusion coefficient can be calculated using (5) and (6):

$$D = \alpha \cdot u \exp(-v^*/v_f), \tag{7}$$

where α is an interatomic distance; u is a thermal atomic rate.

From here, an expression for a viscosity coefficient for a uniform liquid and glass flow can be derived:

$$\eta = Ak_BT/aD, \tag{8}$$

where A is a constant of unity order; k_B is a Boltzmann constant.

The distribution (5) is valid, in one hand, if a local reconstruction of cells occurs so often that gives rise to a metastable equilibrium, on the other hand, a structural reconstruction leading to a crystallization is depressed. Both these conditions, a realization of which could be controlled within a macroscopic description of structure reconstruction theory, may happen to be mutually exclusive. Relations (5) and (6) loose their validity when a temperature is lower than glass-transition one. However, the diffusion and viscosity stay still dependent on a density of free space cells, which are able for a reconstruction under an action of heat fluctuations and stresses.

A thermodynamic approach for a description of amorphous solid containing an excessive free space is one more interpretation of FSM. It is assumed that a solid has cells with and without the free space: a liquid-like and solid-like cell. A state and a concentration of liquid cell is described by a Gibbs distribution, a free atomic energy depending only on v_f. A transition of amorphous solid to a liquid state in this approach, is related to a formation of infinite (leaking through) liquid-like cluster, which appears when a concentration of liquid-like cells reaches a certain critical value (0.15… 0.3).

A polycluster model. A polycluster model, which was introduced and investigated by Prof. Bakai A.S., has no disadvantages of the above considered models. It involves a construction-based determination of topologically disordered structure. A polycluster has a more perfect local order in the most volume fraction in comparison with a system of balls with a stochastic close-packing. At the same time, its topological disorder is stronger in comparison with paracrystals. The fairly perfect local order of polyclusters may go together with a strong topological disorder and with a broken mutual atomic arrangement at finite distances. It is more important that a region with local disorder is only two-dimensional. This is a source of characteristic kinetic and fluctuation properties of polyclusters. Structure defects and a number of polycluster properties have been already described: a mechanism of atomic transfer, mechanical states in a field of external forces and structure fluctuations. And, in spite of the fact that now a polycluster character of real amorphous body structure is not clear, a comparison of theoretical results with experimental data does not indicate serious disadvantages of this model.

An analysis of literature data allows the following conclusions:

- There is no a universal model with an adequate description of structure and properties of amorphous metallic alloy;

- An absence of translation symmetry in an atomic arrangement causes great difficulties in a model construction. A concept of elementary cell can hardly be applicable to an amorphous system. In addition, methods based on a solid interaction with an electromagnetic emission (radiation) have a low efficiency for an amorphous material.

- Offered models can only give only a good description of definite properties but fail to give a full description of amorphous material properties.

A computer modeling plays an important role in a research of amorphous solids. A potential of computer modeling is limited and restrained within a structure containing several thousands of atoms, a simplified description of their interactions for a short time period. However, the computer modeling can be successfully employed for studies of microscopic processes taking place in a volume, which is essentially smaller than a model one, and a relation of these processes to solid macroscopic properties due to a comprehensively full diagnostics. At present, several algorisms for a structure modeling of balls with a random close-packing had been developed. In spite of their difference, they all lead to a formation of solid with similar structure properties. A numerical model is employed for an identification of various structure elements and a statistical description, studies of mechanical properties and a dynamics of amorphous solid. The results of computer modeling are widely applied for an interpretation of various experimental data and yield an information about properties of an amorphous solid.

3.2. PROPERTIES OF AMORPHOUS METALLIC SYSTEM

Recent decades, a solid state physics goes through a vigorous development of direction relating to fabrication and application of an amorphous metallic system *(AMS)*. An amorphous metallic material features so amazing physical and technical properties that their wide application in an engineering can undoubtedly lead to an essentially increased quality and reliability of tools and an essential economy in energy and material resources. An amorphous metallic material can be divided into four main groups:

1. A "transition metal (Fe, Ni, Co)-a metalloid (B, Si, P, C)". Today these materials are most important from a practical viewpoint;

2. A "transition metal (Fe, Ni, Co)-a rare earth metal (Dy, Nd, Gd)";

3. A "transition metal-metal-lanthanide (Sm, Cu, Ho)";

4. A binary and multicomponent material, which is composed of an alkaline-earth and another metal.

At the beginning of 60th, a first amorphous material was obtained using a quenching from a melt. We should underline that this method was employed to fabrication an amorphous alloy in a form of strip or wire, which were separated from a crystalline substrate. A spectrum of their compositions, physical, and technical properties was very wide. Since that time, a quantity off fabricated metallic systems in the amorphous state was constantly increasing.

However, a fast melt quenching and a fabrication of disordered crystal do not exhaust all possibilities of amorphous solid fabrication. An amorphous structure takes part in all possible processes involved in a formation of nonequilibrium structures. A deposition of atomic, ion and plasma flows on substrates, a surface treatment by a glow discharge, a laser, or an electron beam, an irradiation by a neutron or ion flow of high energy under special conditions result in a formation of amorphous structure and are widely used for technological purposed. In most cases, the above methods (this does not concern materials fabricated by a neutron emission) are used to fabricate a relatively thin film. A kinetics of structure formation by the above methods essentially differs from a fast melt quenching, but, probably, there exists a great similarity in a structure and properties of amorphous solids of the same composition, which are obtained by different methods. Unfortunately, today this problem is so poorly understood, that we can hardly present some full comparison.

Mechanical properties. Both for an amorphous metallic alloy (*AMA*) and for a crystalline solid under low deformation, a Hooke's law is valid. An elastic modulus of amorphous alloy is lower than that of a crystalline metal employed as a basis for the alloy. This is related to an excessive free space and reflects a lower average force of interatomic interaction of amorphous states in comparison with crystals. A structure relaxation resulting in a decrease of free space increases an elastic modulus of non-magnetostrictive amorphous alloy by several percent.

In addition to the excessive free space, an essential effect on a value of elastic modules is played by a value and a character of chemical bond, which depends on an alloy composition. An increased amount of metalloid atoms results in a growth of Young modulus from 158 to 187 GPa in Fe-Si-B alloys, from 140 to 152 GPa in Fe-p-C, and from 173 to 175 GPa in Co-Si-B. This indicates an identity in a mechanical behavior of amorphous and crystalline metallic alloy. Like a crystalline metal, *AMA* demonstrates deviations from an elastic behavior in a region of elastic deformation, where the Hooke's law works. Non-elastic phenomena observed under low stresses are the main cause of an internal friction characterizing an irreversible energy loss inside a solid under mechanical oscillations [12-15].

Experimental data on the internal friction for many *AMA*s demonstrated an attenuation maximum, which was present in a temperature dependence at T = 200 to 400K. A height of this maximum depended on an alloy composition and a material structure state. The existence of internal friction peak for many *AMA*s is a proof of atomic configurations with a short-range order characterizing defects of an amorphous structure, which response for an action of external mechanical stress by a local atomic reconstruction. An elementary act of such reconstruction is an atomic jump. A thermal annealing decreases a height of internal friction peak, which seems to be related to a structure relaxation and a decreased defect concentration

Electrical properties of an amorphous metal have two peculiar features: an electric resistance of amorphous metallic alloy at a room temperature is 2 to 4 times higher than that of a corresponding crystalline alloy; a temperature dependence of the amorphous metal, irrespective of its composition, demonstrates a minimum, as a rule, at temperatures below the room one. A higher resistance of amorphous metallic alloy, in comparison to a crystalline one, may be explained as a result of strong action of chemical and configuration disorder on an electron free range. A disorder of atomic arrangement results in that the conductivity electrons of amorphous alloy are scattered more often than it was in their crystalline analogs. The value of electric resistance for the majority of amorphous metals is close to that of liquid of the same composition. In the case of amorphous ferromagnetic materials, a magnetic ordering strongly affects electric properties. A minimum observed at a dependence of resistance on a temperature, may be explained due to a structural disorder and the strong scattering of conductivity electrons, which always exist in an amorphous alloy.

Magnetic properties of an amorphous metal. It was found that an isotropic amorphous structure permits only four types of magnetic ordering: an ordered ferromagnetic; a non-ordered ferromagnetic; a non-ordered antiferromagnetic, and a spin glass. Fig. **2** demonstrates an isotropic amorphous structure of a magnetic material.

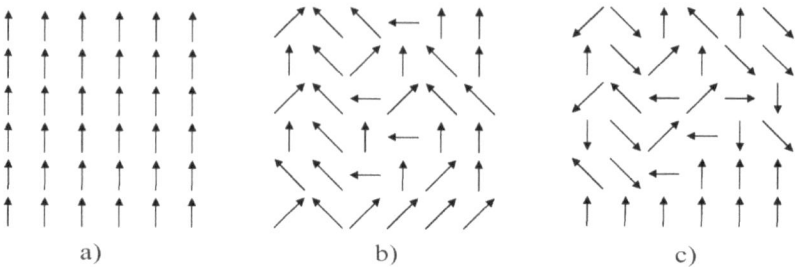

Figure 2: Types of magnetic ordering in an amorphous structure: *a*-for an ordered ferromagnetic; *b*-for a non-ordered ferromagnetic, *c*-for a spin glass.

An existence of the amorphous ferromagnetic material was theoretically confirmed, however, such type of magnetic ordering was not found experimentally in amorphous samples.

Recent decades, a ferromagnetic amorphous alloy (a metallic glass) based on an iron doped with Nb, Cu, Si, B, on Co, and Fe doped with Si and B attracted a special interest. As a result of crystallization, such amorphous alloy transforms into a nanocrystalline material with a grain size 8 to 25 nm and with unique magnetic properties. In a literature, a nanocrystalline alloy Fe-Cu-Nb-Si-B is called a "finemet". It is a magnetic material with a very low coercive force, which is comparable to H_c of an amorphous alloy based on cobalt. It also owns a high magnetic saturation, which is close to that of the iron-based amorphous alloy. The crystallization of amorphous alloys results not only in the formation of magnetically soft, but also a magnetically hard nanocrystalline material with the high coercive force. For example, a fast crystallization during 10 seconds at 750°C in an amorphous soft magnetic alloy $Fe_5Co_{70}Si_{15}B_{10}$ with $H_c = 1$ Am^{-1} transforms into a nanocrystalline alloy with 15 to 50 nm average grain size with $H_c = 8800$ Am^{-1}, and an enhanced residual magnetization. At the same time, a slowly crystallizing of an alloy with the same composition has $Hc = 3300$ Am^{-1} [13-15].

An amorphous semiconducting material is a solid system with a dominating covalent chemical bond, which is characterized by a short-range order (a coordinated arrangement of nearest neighbors) and an absence of long-range order, *i.e.* a translation symmetry at a macro level. A forbidden zone (0.01 to 3.5 eV) is typically present in both an amorphous and a crystalline semiconductor. The forbidden zone is a region of electron energy spectrum, in which a density of states becomes zero (with the exception of a localized state). Active researches of the amorphous semiconducting material started after the Second World War. A first system, which attracted a significant interest of researchers and practical scientists was an amorphous selenium, which took a dominating position as a material applied for fabrication of photosensors, and later-for a xerography. Recently, fundamental and practical researches are directed to an amorphous silicon, which found its application in fabrication of elements for solar batteries, because its cost was essentially lower than that of a single-crystalline silicon (though the latter had a higher efficiency coefficient). An amorphous metallic-silicide resistive alloy also found its practical application.

A non-ordered material (a glass, an amorphous solid, a melt, a polymer, a biological medium, *etc.*) represents an important class of objects. In spite of that chaos, with which usually a structure of glass and amorphous solid (a semiconductor, a dielectric and a metallic material) is associated, there is one universal spatial scale (~ 1nm), which is a parameter playing so important role for a theory like an elementary cell for a crystal. A disorder of non-ordered solids is not absolute. A periodicity of atomic arrangements, which is common for all crystals, remains within the limits of several coordination spheres and is broken in some way in a periphery. The character of broken order differs a glass from an amorphous solid by a form of their structure correlation function. The non-uniformities, which we report about, are not an exotic, individual formation, or an analog of a crystal defect. They are a fragment, which forms the amorphous solid and glass as a whole. A spatial non-uniformity of non-ordered solid with a characteristic ~ 1nm scale gives rise to peculiar features of oscillation properties, a change of electron excitation relaxation mechanism, and determines a specific character of charge transfer.

From the point of view of a fundamental science, a structure of amorphous solid and glass is one of the basic problems for a solid state physics. How this world is organized under conditions of disorder, with which these matters are usually associated? To answer this question is not easier than to answer many fundamental questions of astrophysics and a physics of elementary particle.

Only a crystalline state, which is characterized by a geometric order of atomic arrangement at far distances, stands apart from all matter states. As a result, a great progress was reached in an apprehension of crystalline state by experimental methods and satisfactory theoretical models were derived. However, a non-crystalline material (a great number of amorphous solids finds an application in practice like a glass and a liquid) does not own this order, if they are considered using a "crystallization" criterion. They do not demonstrate a universality of compositions, and every matter has to be studied individually, to construct a model of its structure, and to find some individual signs for a control of its properties. This does not look productive, but is widely spread [12-16, 24-26].

Physical properties and a geometry of amorphous material are closely related. A disorder of amorphous solid has a topological character, and topological defects can hardly be liquidated by low atomic displacements-it needs a global reconstruction of whole structure.

QUESTIONS FOR CONTROL

1. Give a definition of solid amorphous state.

2. What are a short-and a long-range order in a solid?

3. What is a basic difference between the terms of an amorphous and a glass-like state?

4. What is characterized by a radial distribution function?

5. What are the main ways applied to fabrication of an amorphous alloy.

6. What is the main difference between a zone structure of amorphous semiconductor and its crystalline analog?

REFERENCES

[1] Zolotukhin IV, Barmin YuV. Stability and Processes of Relaxation in Metallic Glasses. Moscow: Metallurgy 1991.
[2] Bakai AS. Polycluster Amorphous Solids. Moscow: Energoatomizdat 1989.
[3] Coben MH, Turnbull D. Molecular Transport in Liquids and Glasses. J Chem Phys 1965; 31: 1164-1169.
[4] Turnbull D, Coben MH. On the Free-Volume Model of the Liquid-Glass Transition. J Chem Phys 1970; 6: 3038-3045.
[5] Adler D, Schwartz B, Steel IO. Physical Properties of Amorphous Materials. Plenum Press 1985.
[6] Petrov AA, Gavriliuk AA, Zubitskii SM. Structure and Properties of Non-ordered Solid Systems. Irkutsk: IGU 2004.
[7] Pozdnizkov VA. Physical Material Science of Nanostructured Materials. Moscow: MGIU 2007.

[8] Zolotukhin IV. Physical Properties of Amorphous Metallic Materials. Moscow: Metallurgiia 1986.
[9] Glezer AM, Molotilov BV. Structure and Mechanical Properties of Amorphous Alloys. Moscow: Metallurgiia 1992.
[10] Sudzuki K, Fudzimori H, Hasimoto K. Amorphous Metals. Translated. Moscow: Metallurgiia 1987.
[11] Kunitskii YuA, Korzhik VP, Borisov YuS. Non-crystalline Metallic Materials and Coatings in Engineering. Kiev: Tekhnika 1988.
[12] Felts A. Amorphous and Glass-Like Solids. Moscow: MIR 1986.
[13] Kovernisty YuK. Nanostructured Materials Based on Volume-Amorphous Metallic Alloys. Metals 2001; 5: 19-23.
[14] Yamauchi K, Yoshizawa Y. Recent Development of Nanocrystalline Soft Magnetic Alloys. Nanostruct Mater 1995; 6: 247-262.
[15] Glazer AA. Effect of Fast Crystallization of Amorphous Alloy $Fe_5Co_{70}Si_{15}B_{10}$ on Magnetic Properties. The Physics of Metals and Metallography 1993; 76: 171-178.
[16] Shevchenko SV, Stetsenko NN. Nanostructured States in Metals, Alloys, and Intermetalloid Compounds: Methods of Formation, Structure, Properties. Progress in Physics of Metals 2004; 5: 219-255.
[17] Gopalakrishnan S, Mitra M. Wavelet Methods for Dynamical Problems: With Application to Metallic, Composite, and Nano-Composite Structures. CRC Press 2010.
[18] Johnson WL, Fultz BT. Ravichandran G, Atwater HA. Amorphous metallic foam: synthesis and mechanical properties. California Institute of Technology 2006.
[19] Poon GX, Shiflet SJ, Widom GJ, Ductility improvement of amorphous steels: Roles of shear modulus and electronic structure. Acta Materialia 2008; 56: 88-94.
[20] Schroers J, Paton N Amorphous Metal Alloys Form Like Plastics. Advance Materials Processes 2006; 61-63.
[21] Telford M. The Case for Bulk Metallic Glass. Mater Today 2004; 36-43.
[22] Ashby M, Greer A. Metallic glasses as structural materials. Scripta Materialia 2006; 54: 321-326.
[23] Schroers J, Pham Q, Desai A. Thermoplastic Forming of Bulk Metallic Glass— A Technology for MEMS and Microstructure Fabrication. J Microelectromech Syst 2007 16(2): 240-247.
[24] Books New Nanotechnologies/edit. A. Malik and R.J. Rawat. A.D.Pogrebnjak, A.P.Shpak, V.M.Beresnev. Chapter 2 (p. 25-114). Structure and Properties of Protective Composite Coatings and Modified Surface Prior and After Plasma High Energy Jets. Nova Science Publisher 2009 : 687.
[25] Pogrebnjak AD, Lozovan AA, Kirik GV, *et al*. Structure and Properties of nanocomposite, hybrid and polymers coatings,Publ. House URSS, Moscow, 2011, 344.
[26] Azarenkov NA, Beresnev VM, Pogrebnjak AD, *et al*. Fundamentals of Fabricated Nanostructured Coatings, Nanomaterials and Their Properties, Publ. House URSS, Moscow 2012; 352.

CHAPTER 4

Fulerene, Fulerite and Nanotubes

Abstract: A definition of fullerene, fullerite, and nanotube is presented. Various models, forms, and a geometry of nanotubes (a one-dimension and a three dimension image) are considered. A definition of nanotube chirality and polygonization is presented.

Keywords: Fulerens, fulerites, nanotubes, CNT.

4.1. INTRODUCTION

A carbon is a widely abundant element. In nature, its solid state is met in a form of graphite and diamond. Such artificial modifications of carbon like a carbyne and a lonsdaleite were fabricated. The lonsdaleite was revealed as a meteorite component. In 1985, a group of researchers-Kroto H. W., Heath J. R., O'Brien S. C., Richard Smalley, Robert F. Curl Jr. studied a mass-spectrum of graphite vapor formed by a laser ablation and found a peak of maximum amplitude corresponding to a cluster, which contained 60 to 70 carbon atoms (Fig. **1**) [1-3, 24-31].

A subsequent study of these formations demonstrated that among all known compounds, the most stable one is a molecule with an even number of atoms, first of all, containing 60 to 70 atoms of C_{60} and C_{70}. The compound C_{60} has a spherical form looking like a football, and C_{70} resembles a melon (Fig. **2**).

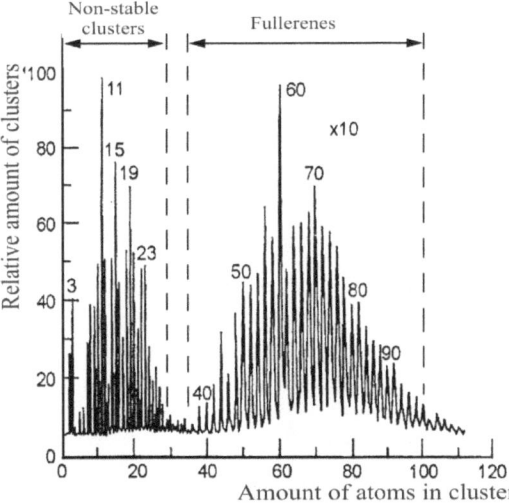

Figure 1: A mass-spectrum of carbon cluster, which was formed in a graphite by a laser ablation [1].

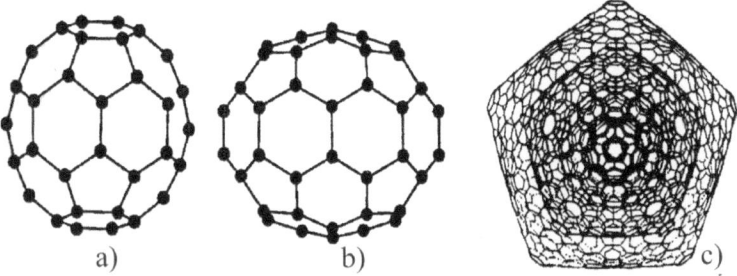

Figure 2: A fullerene molecule: *a*-a C_{60} molecule; *b*-a C_{70} molecule, *c*-a predicted fullerene molecule containing more than 100 carbon atoms [3].

Carbon atoms are arranged in a sphere surface at a vertex of pentagons and hexagons. The molecule looks like a football with 12 black pentagons and 20 white hexagons. The C_{60} molecule is able to crystallize forming a cubic lattice. A polyhedral carbon cluster was named a fullerene, and the most widely abundant molecule C_{60} was named as a Buchminster fullerene after an architect Buckminster Fuller, who constructed a dome for an USA pavilion for the Montreal exhibition 1967 looking like junctions of pentagons and hexagons. As an architect, Mr. Fuller proposed a construction looking like a polyhedron spheroid, which could serve as a ceiling covering a big area, and as a mathematician, Mr. Fuller applied a systematic approach for an analysis of structures with different origins and demonstrated that a structure is a self-stabilizing system. However, to tell the truth, this form already existed among the semi-regular forms of Archimedes. In addition, a figure of wooden model of such form, which was fabricated by Leonardo da Vinci, was saved, and Leonhard Euler derived a formula, which allowed a calculation of polygon number for various surfaces.

A fullerene is the fourth allotropic form of carbon (the former three are a diamond, a graphite, and a carbide). Further, for a good accuracy, we shall call the solid phase C_{60} like a fullerene, and an individual molecule C_{60} like a fullerene molecule. The molecule C_{60} contains fragments with a five-fold symmetry (pentagons), which are forbidden for a non-organic molecule in nature. In this connection, we have to agree that the fullerene molecule is an organic matter, and the fullerene itself represents a molecular crystal serving as a joining element between an organic and a non-organic matter.

One should note that a discovery of fullerene has its own history: an assumption that the high-symmetry carbon molecule resembling a football existed, for the first time, was predicted by Japanese scientists E. Osawa and Z. Yoshida in 1970. Later, Russian scientists D.A.Bochvar and E.G.Galperin performed a first theoretical quantum-chemical calculation for this molecule and proved its stability. In 1985, English scientists Walter Kroto with colleagues [4] synthesized the C_{60} molecule. For this purpose, a solid graphite target was subjected to a powerful laser radiation. As a result, a chaotic plasma of 5000 to 10 000°C temperature was formed. In this plasma, the molecule C_{60} was synthesized and identified using a mass-spectrometry device, which allowed atoms and molecules to be sorted according to their mass. The formation of fullerene molecule from the plasma represents an excellent example of how to organize a structure ordering from a chaos, that now, it is one of the most interesting and thrilling fields in a modern science! As we have already mentioned, the molecules C_{60} are ordered in a space under definite conditions and arrange inside a narrow crystal lattice, or, as they say, the fullerene forms a crystal. In order that the C_{60} molecules were regularly arranged in the space, they must be bound to each other like a molecule atom. Binds between the fullerene molecules are weak. They are called Van der Waals' binds in an honor of Holland scientist.

The fullerene is known by an unusual crystallographic symmetry and unique properties [5]. All its covalent bonds are saturated, therefore, an individual molecule can interact with another one only due to a weak Van der Waals force. These bonds arise because a negative electron charge and a positive nucleus charge of electrically neutral molecule are spatially separated. As a result, the molecules polarize each other, *i.e.* shift a center of positive and negative charge in a space and induce their interaction. This interaction is sufficient to form a crystalline structure of spherical molecules. Such material is called a *fullerite*. A stable molecule is characterized by a chain configuration, which is formed by a five-and six-member ring.

In most cases, their carbon atom has three spatial bonds (similar to a diamond lattice). A length and an angle of these bonds are also similar to the diamond structure.

Later, it was revealed that a natural fullerene also exists. In 1992, the fullerene was found in a natural carbon mineral schungite (the mineral was named after a small village Schunga in Karelia Region, Russia). To tell the truth, an amount of fullerene content in the schungite is low, it does not exceed 10 to 3 %. In 1993, another multi-atomic molecule and a carbon microparticle C_{70} (a nanotube, a nest-doll, a bulb) were found in schungites.

A classic way to fabricate a fullerene is an vaporization of carbon in a vacuum resulting in an overheating of carbon vapor to 10^4 K [6]. This overheated vapor is intensively cooled in a jet of inert gas (for example, a helium). As a result, a powder with a significant amount of two groups of clusters (molecules) (of a small size with an odd number of carbon atoms (to C_{25}) and of a big size with an even number of carbon atoms (C_{60} and C_{70})) is deposited. They can be separated using methods applied in a powder metallurgy. Moreover, clusters referring to the first group are unstable. A formation of molecules with a high amount of atoms (C_{100} and more) is also possible by a selection of process parameters. There is a number of other methods of fullerene fabrication [7, 32-34].

Today, people are able to fabricate a doped fullerene by doping different atoms and molecules to their molecules including an introduction of doping element atoms into an inside molecule volume. Two fullerene molecules can be joined into a dimer or an initial monomer structure can be subjected to a polymerization due to an action of high pressure or laser irradiation. A method of vacuum thermal deposition of a blend of desired composition to a substrate, for example GaAs (Fig. 3), is employed to fabricate a thin composite film of 200 to 600 nm thickness based on a fullerene matrix [8, 24-27, 35]. A blend of C_{60} powder of 99.98 % purity and CdTe was prepared by their simultaneous comminuting to 1μm size and sintering at 300°C temperature.

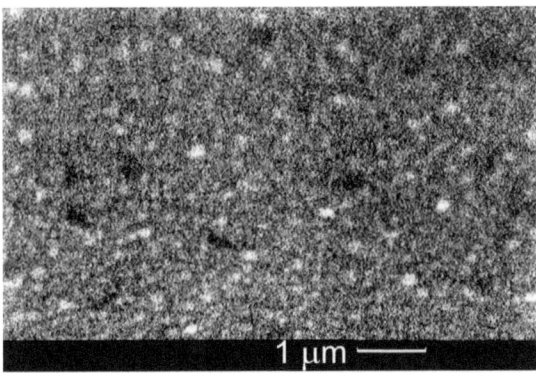

Figure 3: A surface of "fullerene C_{60} 40%CdTe" film [8].

The deposition was performed in a vacuum of 10^{-6} Torr pressure and at about 160 °C substrate temperature. A resulting film did not have a notable spatial non-uniformity of chemical composition.

A very high fullerene hardness successfully employed for a fabrication of fullerite micro-and nanotools, which are applied for a treatment and a testing of superhard materials including a diamond. For example, a fullerite pyramid made of C_{60} can be applied in an atomic force microscope for diamond and diamond film hardness measurements. The fullerene is a promising material for an electronics and an optics [9]. The fullerene and the compound based on it have also a high potential for a nanostructure formation. A fullerene film may be applied tofabrication a two-dimension photon crystal. Optical properties of fullerene films may be changed by a semiconductor material doping, for example, CdSe and CdTe [8].

In 1991, a report of scientists from the Bell Laboratory (USA) attracted a great interest. They reported that a potassium-doped fullerene was a superconductor and a temperature of its transition to about to a superconducting state was 18K [3]. Later, it was found that a fullerite, which was formed on the base of other alkaline (excluding a sodium) and alkaline-earth metal, was also a superconductor. A maximum transition temperature turned to be 42K, *i.e.* a certain fraction of metallic fullerenes turned out to be a high-temperature semiconductor. In 1994, the authors of works dealing with an identification of superconducting carbon-containing phase, which was found in a schungite, discovered a metallic fullerene Cu_nC_{60}. Its transition temperature was higher than that of a liquid nitrogen.

Another interesting property of the doped fullerene is a ferromagnetism discovered in 1991 [10]. That time, a soft organic ferromagnetic material C_{60}-*TDAE* was fabricated (*TDAE* was a tetradimethylaminoethylene with a Courier point T_C = 16 K). In 1992, a ferromagnetic with T_C = 30 K based on a fullerene doped by an iodine and a bromine was formed [10-14].

A discovery of unique carbon structures and researches of their properties have a continuation together with an attempt to understand how a fullerene can be applied in an electronics, a biology, a medicine, and other fields of human activity. However, even now it is evident that the fullerene is a bridge joining a non-organic and an organic substance, an alive and a non-alive matter. An this seems to serve a serious reason for wide-scale researches of fullerene and fullerite properties in scientific laboratories allover the world. After the discovery of C_{60} and C_{70} fullerenes, which were found in graphite combustion products after an action of electrical arc or high power laser beam, a particle containing a carbon atom of regular form and with a size ranging from several ten to several hundred of a micrometer was revealed. It was called a nanoparticle.

A question arises, why the fullerene was not discovered earlier, if it could be formed from such a widespread material as a graphite? There are two main reasons: the first one is that a covalent bond of carbon atom is very strong. To break it, one needs a temperature above 4000°C. Second, to find them, one needs a very complicated equipment-a scanning electron microscope with a high resolution. Now, it is known that a nanoparticle can have a very fantastic form. A nanotube is the most interesting object for a nanoelectronics, which is now replacing a microelectronics.

A nanotube. The first nanotube was revealed in 1991 by a Japanese scientist Sumio Iijima in the course of researches of a carbon electrode surface. A carbon electrode was used in an electric arc discharge device, which was applied for a fullerene fabricating. The first fullerene looked like a multi-wall carbon nanotube (*MWCNT*) formed from a pair or several ten of concentric cylinders placed around a common central hole with an interlayer distance, which was close to that of a graphite (0.34nm). Its internal diameter changed from 0.4nm to several nm, and its external diameter usually ranged from 2 nm to 20-30 nm depending on the number of layers. The edges of *MWCNT* are usually closed by an inset of pentagon defects embedded into a graphite network. A length of such *CNT* (a carbon nanotube) varied from 1μm to several centimeter.

The nanotube is a long cylinder formed by a six-angular honeycomb carbon lattice, which edges end with two fullerene fragments (Fig. **4**).

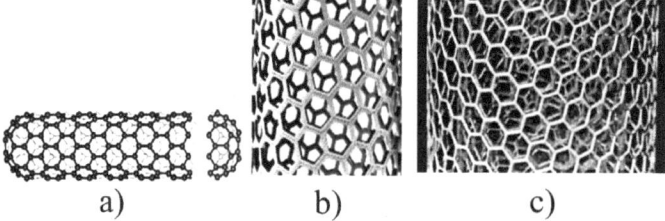

Figure 4: A schematic image of a carbon nanotube (CNT): *a*-a *CNT*; *b*-a singlewall *CNT*; *c*-a multiwall *CNT*.

The nanotube diameter depends on a size of semi-fullerenes forming the structure edges. The first information about a single-wall carbon nanotube (*SWCNT*) was reported in 1993. It was presented as a seamless cylinder made of one graphene sheet. Its diameter varied from 0.4nm to 2-3 nm, its length usually did no exceed a micrometer range. The *SWCNT* can often join with each other and form a bundle (a rope). The bundle is arranged at hexagon angles and forms a crystalline structure. The *SWCNT* may be considered as a cut out strip of a graphite sheet rolled up as a tube (Fig. **5a**). A diagram is reproduced for *(n, m)* = *(4.2)*. The diameter and a helicity of *SWCNT* are unambiguously characterized by a convolution vector $\vec{C_h} = n\vec{a_1} + m\vec{a_2} = (n,m)$, which unites crystallographically equivalent regions in a two-dimensional (2D) graphene sheet, where $\vec{a_1}$ and $\vec{a_2}$ are the vectors of graphite lattice, and *n, m* are the integral numbers.

Extreme cases of achirality (a zigzag-configuration (n, 0) and a "saddle"-configuration (n, n)) are presented in Fig. **5b** by dashed lines. A translation vector $\vec{T}$ is parallel to a tube axis and is orthogonal to $\vec{C_h}$, its value represents a tube single cell length *(n, m)*. A rolled up area with $\vec{T}$ and $\vec{C_h}$ vectors (Fig. **5b**) corresponds to a repeated tube cell *(n, m)*. Consequently, *(n, m)*, which is a nanotube symmetry, determines a single cell dimension highly varying from one tube to another [15-20, 24-27].

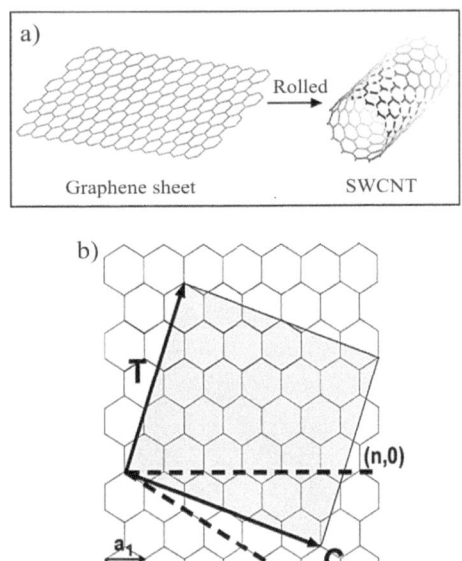

Figure 5: A single-wall *CNT*: *a*-a formation of infinite graphene sheet (a layer) rolled-up like a tube; *b*-a schematic presentation of two-dimensional graphene sheet.

A calculation of electron energy-band structure predicts that *(n, m)*-indexes determine whether *SWCNT* would be a metal or a semiconductor. The nanotube with a corresponding translation index of *(n, 0)*-or *(n, m)*-type has only one reflection plane, and consequently, only two screwing symmetry. All other nanotubes are characterized by three equivalent screwing symmetries. In general, the nanotube of *(n, 0)*-type is called a zigzag nanotube (for example, a nanotube (8.0)), an *(n, n)*-type-a "saddle" nanotube (for example, a nanotube (10, 10)), Fig. **6** [20].

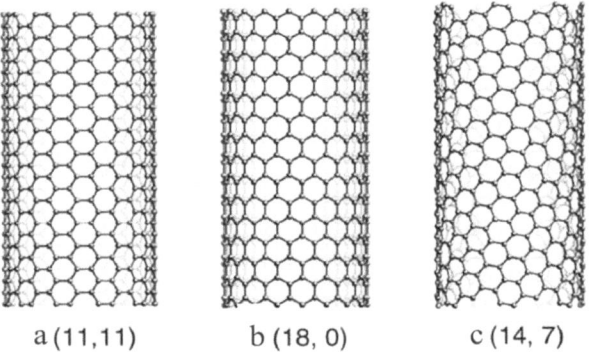

Figure 6: A computer image of single-wall carbon nanotube: *a*-a "saddle"-type; *b*-a zigzag-type; *c*-a chirality-type.

Chirality numbers *m* and *n* seem to be definitely related to the *SWCNT* diameter *D*:

$$D = \sqrt{m^2 + n^2 + mn} \frac{\sqrt{3}d_0}{\pi}, \tag{1}$$

where: $d_0 = 0.142$ nm is a bond length *c-c* in a graphite plane. On the other hand, a chirality angle and chirality numbers for *SWCNT* are related to each other as:

$$\sin \alpha = \frac{3m}{2\sqrt{n^2 = m^2 + nm}} \text{ or } \alpha = \tan^{-1}\left[\sqrt{3n}/(2m+n)\right], \tag{2}$$

where α is limited by $0 \leq \alpha \leq \frac{\pi}{6}$, according to a geometric symmetry of hexagonal network, α = 0° for the "saddle"-type nanotube and α = 30° for the zigzag-configuration. The nanotube axis is positioned along one straight line with two *c-c* bonds of hexagon for the (10, 10) "saddle"-configuration nanotube.

Fig. 7 shows an image of 10 x 10 nm^2 topography for an isolated *SWCNT* taken in three dimensions by a scanning tunneling microscope. The nanotube diameter may be considered as a structure of 0.98 ± 0.03 nm. A simultaneous atomic resolution of both the nanotube and Si substrate indicates a very low contamination level. The image simultaneously demonstrates the details of silicon dimers and a carbon lattice chirality.

Figure 7: TA three-dimension presentation of CNT topography of 10 x 10 nm^2 area for a single-wall carbon nanotube, which was preliminarily prepared under high vacuum conditions and then physically sorbitized to Si(100)$^{-2}$ x 1: an H surface.

However, when high-resolution experimental data about structure became accessible, the *CNT* did not look so ideal, as it was earlier. Defects with 5 to 7 atomic rings, loops, compositions, Stone-Wales defects, and impurities were found in the fabricated CNT. Very interesting structure properties were found near the edges of all tubes, in a region where a graphite cylinder was closed. These interesting structure properties arose because topological pentagon-type defects were united into a hexagonal carbon lattice. Various complexes and structures (for example, conical sharp peaks of definite forms) arose, depending on the way of pentagon distribution around the tube edges, which continued until these edges were fully closed. In a theory, the tube edges might have various electron structures depending on a presence of topological defects. And this was also confirmed experimentally. An electron structure arising at the *CNT* edge due to a presence of defects seems to be important for a number of reasons. For example, properties of a nanotube field emission may be strongly dependent on a presence of localized resonance state. A presence of defects like a pentagon-heptagon pair (a distant defect pair forming a step in a surface) gives an attractive possibility to change a curvature and a helicity without substantial bond distortions. There is an information about a nanotube, a chirality of which alters along its axis [21-23, 26-32].

There are also several types of defects, which lie out of a cylinder plane of carbon nanotube. In most cases, they are boundary dislocations, which are formed due to a chirality of individual layer and a subsequent rotation disorder arising between the layers. Due to the chirality and the disorder, atoms of neighboring layers get chaotic. Consequently, in a general case, the structures do not demonstrate a three-dimension ordering, which is observed in a subsequently packed single crystalline graphite structure. SEM images also show a presence of boundary dislocations, scrolled layers, and edge Frank dislocations, which are distributed along the tube axis [36-39]. In addition to a perfect cylindrical tube, a polygonized tube was revealed. An interval between cylinders in every individual tube deserves a special consideration. A real spatial analysis of images demonstrates that a layer interface of a multi-wall nanotube varies within 0.34 to 0.39 nm. Generally, the interval between cylinders increased with a decreasing diameter of graphite cylinder due to an increasing bending of graphite layers.

QUESTIONS FOR CONTROL

1. What is a fullerene? What is a difference between C_{60} and C_{70}?

2. What are the fullerene features?

3. What is a fullerite?

4. What are the structure features of a single-wall and a multi-wall *CNT* in comparison with the other carbon structure states (a diamond, a graphite, a fullerene).

5. Determine a relation between a single-wall *CNT* diameter and the chirality numbers *m* and *n*, as well as between a chirality angle and the chirality numbers.

6. What are the features of *CNT* with a "saddle"-and a zigzag-type configuration?

REFERENCES

[1] Eletskii AV, Smirnov BM. Fullerenes. Physics-Uspekhi 1993; 163: 33-60.
[2] Eletskii AV, Smirnov BM. Fullerenes and Carbon Structure.Physics-Uspekhi 1995; 165: 977-1028.
[3] Smolin RE. Discovering Fullerenes. Physics-Uspekhi 1998; 168: 323-329.
[4] Kroto HW. C60: Buckmisterfullerene. Nature 1985; 318: 162-167.
[5] Kratschmer W, Lamb LD, Fostiroponlos K, Hoffman DR. Solid C60: a New Form of Carbon. Nature 1990; 347: 354-362.
[6] Hebard AF, Rosseinsky MJ, et al. Superconductivity at 18K in Potassium-Doped. Nature 1991; 350: 600-607.
[7] Allemand PM, Khemani KC, et al. Organic Molecular Soft Ferromagnetism in Fullerene C60. Science 1991; 253: 301-310.
[8] Rao CN, Ram S. Phase Transitions, Superconductivity and Ferromagnetism in Fullerene Systems. MRS Bull 1994; 19: 28-34.
[9] Kozyrev SV, Rotkin VV. Fullerenes: Structure, Dynamics of Crystal Lattice, Electron Structure. Semiconductors 1993; 27: 1409-1413.
[10] Osawa E, Yoshida M, Fujita M. Shape and Fantasy Fullerenes. MRS Bull 1994; 19: 33-42.
[11] Dikii VV, Kabo GJa. Thermo-Dynamical Properties of C60 and C70 Fullerenes. Russ Chem Rev 2000; 69(2): 107-117.
[12] Kolodney E, Tsipinyuk B, Budrevich A. The Thermal Stability and Fragmentation of C60 Molecule up to 2000K on the Millisecond Time Scale. J Chem Phys 1994; 100: 8542-8559.
[13] Lozovik YuE, Popov AM. Formation and Growth of Carbon Nanostructures-Fullerenes, Nanoparticles, Nanotubes, and Cones. Physics-Uspekhi 1997; 167: 751-774.
[14] An KH, Jeon KK, Moon J.-M, et al. Transformation of Single-Walled Carbon Nanotubes to Multi-Walled Carbon Nanotubes and Onion-like Structures by Nitric Acid Treatment. Synthesis Metals 2004; 140: 1-8.
[15] Buria AI, Arlamova NT, Kholodilov OV, Sytnik SV. Investigation of Thermal destruction of Phenilon and Based on it Carbon Plastics. Mechanics Of Materials And Constructions 2001; 6: 58-61.
[16] Tkachiov AG, Zolotukhin IV. Devices and Methods for Synthesis of Solid Nanostructures. Moscow: Mashinostroenie 2007.
[17] Makarova TL, Zakharova IB. Electron Structure of Fullerenes and Fullerites. St.-Peterburg: Nauka 2001.
[18] Eletskii AV. Carbon Nanotubes and Their Emission Properties.Physics-Uspekhi 2002; 172: 401-438.
[19] Eletskii AV. Sorption Properties of Carbon Nanostructures.Physics-Uspekhi 2004; 174: 1192-1231.
[20] Komarov FF, Mironov AM. Carbon Nanotubes the Present and Future. Phys Chem Solids 2004; 5: 411-429.
[21] Odom TW, Huang J.-L, Kim P, Lieber ChM. Structure and Electronic Properties of Carbon Nanotubes. J Phys Chem 2000; B104: 2794-2809.
[22] Gorelik OP, Diuzhev GA, Novikov DV, et al. Cluster Structure of Particles of Fullere-Containing Black and Fullerene Powder C60. Journal T.F 2000; 70: 118-125.
[23] Albrecht PM, Luding JW. Ultrahigh-Vacuum Scanning Tinneling Microscopy and Spectroscopy of Single-Walled Carbon Nanotubes on Hydrogen-Passivated Si(100) Surfaces. Appl Phys Lett 1992; 83: 5029-5031.

[24] Advani G. Suresh. Processing and properties of nanocomposites. World Scientific Publishing Co. Pte. Ltd. 2007.

[25] Dresselhaus MS, Dresselhaus G, Avouris Ph. Carbon nanotubes Berlin: Springer 2001.

[26] Cataldo F, Milani P, Da Ros T. Medicinal Chemistry and Pharmacological Potential of Fullerenes and Carbon Nanotubes. Springer Science and Business Media 2008; vol.1.

[27] Goldberg-Oppenheimer P, Eder D, Steiner U. Carbon Nanotubes Alignment *via* Electrohydrodynamic Patterning of Nanocomposites. University of Cambridge: Dep. Materials Science and Metallurgy 2010.

[28] Reich S, Thornsen C, Maultzsch J. Carbon nanotubes. WILEY-VCH Verlag GmbH & CO. KGaA 2004.

[29] Seminario J.M. Molecular and Nano Electronics. Analysis, Design and Simulation Elsevier 2007.

[30] Shimizu T. Self-Assembled Nanomaterials II: Nanotubes Springer, 2008.

[31] Tomanek D., Enbody R.J. Science and Application of Nanotubes, NY:. Kluwer academic publishers, 2002.

[32] Beresnev VM, Pogrebnyak AD, Turbin PV, *et al.* Tribotechnical and Mechanical Properties of Ti-Al-N Nanocomposite Coatings Deposited by the Ion-Plasma Method. Journal of Friction and Wear 2010; 31 (5): 349-355.

[33] Pogrebnjak AD, Uglov VV, Il'yashenko MV, *et al.* Nano-microcomposite and combined coatings on Ti-Si-N/WC-Co-Cr/steel and Ti-Si-N/(Cr3C2)75-(NiCr)25 Base: Their structure and properties. Ceramic Engineering and Science Proceedings 2010; 31 (7): 115-126.

[34] Pogrebnjak AD, Sobol OV, Beresnev VM, *et al.* Phase Composition, Thermal Stability, Physical and Mechanical Properties of Superhard On Base Zr-Ti-Si-N Nanocomposite Coatings Nanostructured Materials and Nanotechnology IV: Ceramic Engineering and Science Proceedings 2010; 31(7): 127-138.

[35] Beresnev VM, Sobol' OV, Pogrebnjak AD. Thermal stability of the phase composition, structure, and stressed state of ion-plasma condensates in the Zr-Ti-Si-N system. Tech Phys 2010; 55 (6): 871-873.

[36] Beresnev VM, Sobol' OV, Pogrebnjak AD, *et al.* Features of the structurally-phase state of multicomponent coatings on basis of Zr-Ti-Si-N system formed by the method of the vacuum-arc deposition). Problems of Atomic Science and Technology 2009; 6: 158-161.

[37] Pogrebnyak AD, Danilenok MM, Drobyshevskaya AA, *et al.* Investigation of the structure and physicochemical properties of combined nanocomposite coatings based on Ti-N-Cr/Ni-Cr-B-Si-Fe. Russian Physics Journal 2009; 52 (12): 1317-1324.

[38] Cherenda NN, Uglov VV, Poluyanova MG, *et al.* The influence of the coating thickness on the phase and element composition of a 'Ti coating/steel' system surface layer treated by a compression plasma flow. Plasma Processes and Polymers 2009; 6 (SUPPL. 1): S178-S182.

[39] Pogrebnjak AD, Lozovan AA, Kirik GV, *et al.* Structure and Properties of nanocomposite, hybrid and polymers coatings,Publ. House URSS, Moscow, 2011, 344.

CHAPTER 5

Nanocomposite Material

Abstract: A nanocomposite material is considered as a class. Definition and classification of these materials on the basis of their geometrical dimension are presented. A polymer composite is described. The nanocomposite properties are considered. A nanosized metallic-polymer (with and without pores) is briefly described. A principal scheme of nanoparticle formation in a grafted layer is considered.

Keywords: Types of nanocomposites, nanocomposite properties.

5.1. NANOCOMPOSITE

A composite material is a heterogeneous system, which contains, at least, one phase with a size of structure element lower than 100nm, *i.e.* it should be considered as a nanocomposite material. Such material manages to combine the best properties in spite of a small size of its structural element: the improved physical-mechanical, chemical, magnetic characteristics, a high temperature resistance, a stable nanostructure, which remains stable in the process of its fabrication and exploitation. Fig. **1** shows how the mechanical properties of modern materials will tend to be changed in future.

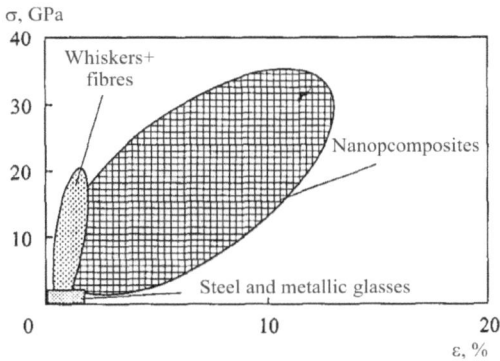

Figure 1: The trends of modern material future development [1].

Today, scores of nanocomposite materials combining high physical-mechanical properties had been already fabricated. K Niihara was the first, who classified a nanocomposite material according to a geometrical size of matrix grain and a second phase particle [2]. An already fabricated and described nanocomposite is not exactly a nanocomposite material but a micronanocomposite material, a matrix and various inclusions of which have a nanosize range. A new classification, in which a matrix is assumed as a nanocrystal and a second phase has a varying dispersion and morphology (Fig. **2**), seems to be necessary for a nanocomposite material [2].

The nanocomposite deserves a special consideration, since there is a great number of scientific publications, and monographs concerning the formation and studies of its physical, mechanical, and chemical properties. In the Chapter 8 entitled "Structure and Properties of Nanostructured Film and Coating", we shall consider, in details, certain physical-mechanical and thermal-physical properties of the nanocomposite material fabricated in the form of a nanocrystalline coating. In this Chapter, we briefly concern the properties of a nanocomposite material with a polymer matrix and a metal nanoparticle employed as a second phase.

A nanopolimer Composite. A nanopolymer material science is a new trend in a polymer composite material fabrication. The properties of the whole composite material cannot surpass those of its individual phases or interphase layers. A nanoparticle is used as a second phase to form the polymer nanocomposite material, and its size and surface topology should be taken into account in this process (Fig. **3**).

Alexander D. Pogrebnjak and Vyacheslav M. Beresnev
All rights reserved-© 2012 Bentham Science Publishers

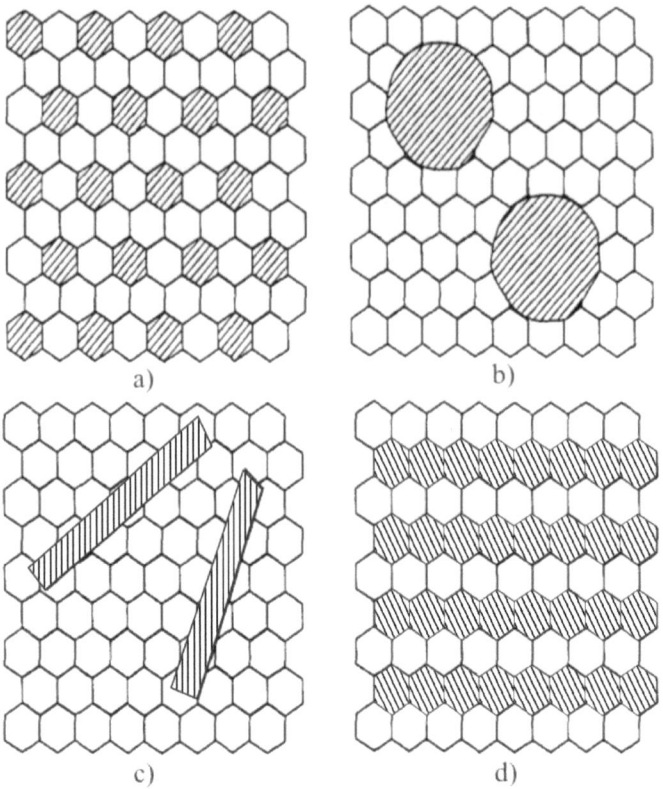

Figure 2: Types of a nanocomposite material [2, 3]: *a.* a nano-nano; *b.* a nano-micro; *c.* a nano-whisker; *d.* a nano-nanolayer.

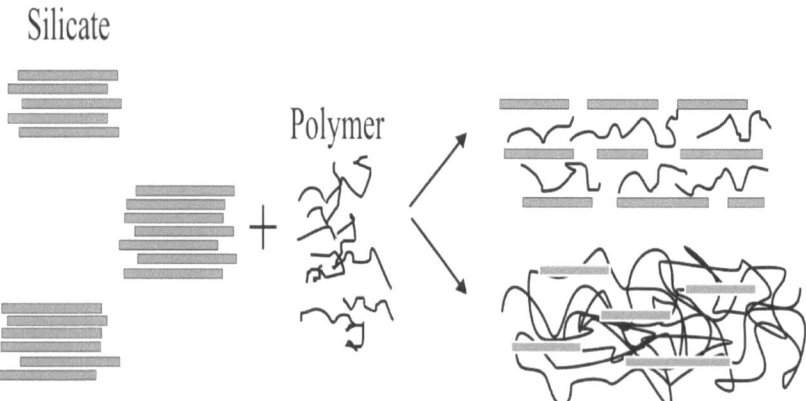

Figure 3: A layered nanocomposite material based on an aluminum silicate and a polymer when its content is low and high [4].

The greatest progress in the nanocomposite material formation was achieved with the help of a sol-gel technology, in which an alcoholate of a certain chemical element and an organic oligomere served as an initial component.

Many compounds (a polystyrene, a polyimide, a polybutadiene, and a polymethylmetacrylate) may be used as the organic component, and depending on reaction conditions and a content of components, materials with various types of a supramolecular organization can be formed [5-9].

The nanocomposite material based on a ceramics and polymer combines all qualities of these individual components: a flexibility, an elasticity, a good processing ability of polymers, as well as a high hardness, a high wear resistance, and a high refractive index of glasses. Following this way, many material properties can be improved in comparison with those of initial components.

A layered nanocomposite material is based on a ceramics and a polymer. A natural nonorganic structure is used for its formation, such as a montmorillonite or a vermiculite, which occurs, say, in clays. In the course of ion interchange, the montmorillonite layer of about 1nm thickness is saturated by a monomer precursor with an active finite order group. Then, a polymerization is performed.

In comparison with a pure polyimide, a moisture permeability of a polyimide nanocomposite polyimide containing only 2 wt.% of a silicate decreased by 60% and a thermal expansion coefficient-by 25%. The main problem of the layered nanocomposite based on the ceramics is a uniform distribution of a layered structure and a monomer inside the material.

In is known that a metal and semiconductor nanocluster, which is employed as a second phase in the polymer nanocomposite material, may be formed in different ways: by a metal evaporation, and sputtering, a reduction from their salts, or other ways. For example, a silver, a gold, or a palladium cluster of 1 to 15 nm size is dispersed to a polystyrene film using a polymerization of liquid monomer, in which a metal was preliminarily deposited from vapor. The metallic clusters were joining into an agglomerates, the sizes of which varied and reached several ten of a nanometer. A composite film, which was formed by a simultaneous metallic vapor-deposition or a benzol and a hexamethyldisilazane (HMDS) polymerization from a plasma, had the same structure [10-12, 14, 15].

The nanocomposite material can be also formed by the simultaneous metal vapor-and active precursor (paracyclophane) deposition and the subsequent polymerization. The paracyclophane molecule, which passes through a pyrolysis zone ~ 600 °C, is transformed into an active intermedium component, which is deposited to a cool substrate together with the metal atoms or the semiconductor molecules. Then, a poly-p-xylene (or its derivatives) is formed in the course of thermal-or photo-polymerization, and a nonorganic nanoparticle or cluster of 1 to 20 nm size (depending on the precursor chemical structure and the polymerization conditions) appears in the polymer matrix. Mainly, the particles are localized in a polymer amorphous region and form a superlattice [17-21].

A thin film, which was formed using the above mentioned way, contained the atoms of various metals and other matters (for example, fullerene C_{60}), its component concentration can be easily varied, and the resulting nanocomposite materials have a high purity. The nanocomposite based on metals, semiconductors, or the poly-p-xylene owns an unusual photo-physical, magnetic, catalytic, and sensor property. When the metal concentration is low, the nanoparticle does not interact with another one, since it is separated by a matrix. In this case, the film resistance is maximum ~ 10^{12} Ohm. When the metal concentration was increased, the resistance decreased to 100 Ohm.

Also, the nanocomposite material can be based on a blocked copolymer. When the latter are joined with each other, they form a block or a domain, which is repeated many times in a polymer chain.

A magnetic property of the polymer nanocomposite material with a surface oxide film, which contains an iron particle (~20 nm), has high value of coercitive force (20800 A/m).

Fig. 4 shows a dependence of magnetization on a magnetic field inductivity for Fe particle, which was coated by a polymer.

A sharp increase of magnetization, which is observed near a zero field value, seems to be of interest. The dependence is observed only when a particle has the polymer coating. The polymer matrix influences a magnetic interaction and an anisotropy, a encapsulating, however, an individual nanoparticle. A technology, which is applied to form such materials, involves an introduction of pentacarbonyle iron into the styrene monomer and a treatment by a microwave plasma. An average particle size is 15 to 20 nm.

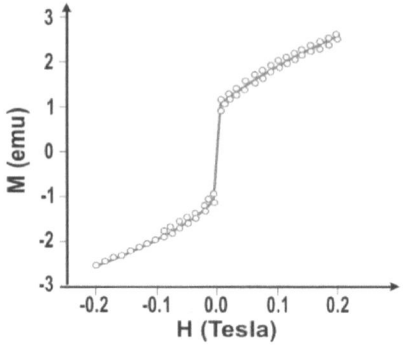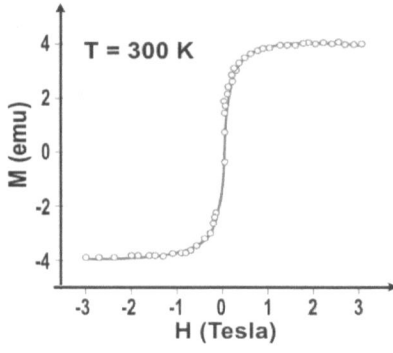

Figure 4: A dependence of magnetization on a field value [8].

It is known, that the polymer interacts with a nanoparticle in two principally different ways-physically or chemically. A non-covalent nanoparticle interaction with a macromolecule is extremely weak (of 10^{-4} J/m^2 order). An efficiency of such interaction under conditions of chemisorption is determined by a quantity of polymer polar groups adsorbed per a surface unit independently on a form of macromolecule. Not only a presence of definite functional group in the polymer is important, but also its intensive interaction with surface atoms of a nanoparticle. A formation of nanocomposite based on an insoluble polymer in a very complicated, because a reduced ion diffuses into a polymer matrix. The nanoparticles of reduced metal are localized in the matrix pores due to a *counterflow diffusion* following the following succession of stages: a penetration of metal ions and reduced agents into a polymer matrix, a diffusion of reagents into the matrix volume, and a chemical reaction. A size of generated nanoparticle depends on the interaction conditions and the parameters of polymer pore structure, and only a little on a metal nature. An increase of metal content is reached mainly due to an increase of particle size rather than an increase of particle quantity. The structure of such nanocomposite and the distribution of metallic layers along the polymer matrix cross-section are determined by a reaction zone width, which depends on a ratio between a diffusion coefficient D and a chemical reaction constant k. When $D \ll k$, a deposition rate of metallic particle is limited by the diffusion, and the reaction zone width is minimum. When $D \gg k$, the recreation zone stretches all over the total cross-section of polymer film. A control of these parameters (a solution viscosity, a temperature, a reagent concentration, *etc.*) allows a fabrication of various modifications of nanocomposite materials. Various chemical variations of nanocomposite materials may be formed depending on a nature of polymer matrix under a reduction of metal ions. For example, a copper oxide is formed when Cu^{+2} is reduced in a swelling matrix (a polyvinyl spirit, a cellulose, *etc.*), and a copper is mainly formed in a porous matrix (a polyethylene, a polytetrafluoroethylene) [23, 24].

One of the ways to form a metal propylene is a high rate thermal decomposition of precursors (in most cases a metal carbonyl) in a polymer solution melt. The melt saves a short-range order of initial polymer structure, and its voids become accessible for a localization of forming particles. Initially, they penetrate into interspherulite regions of the polymer matrix, into a volume between lamella, and into spherulite centers. A strong interaction between nanoparticles and polymer chains is observed.

Products formed by a reduction of metal ions in a polymer nanopore as in a nanoreactor, for example, an ion-exchange resin, are also nanoheterogeneous composite materials. There, a pore plays a role of transportation way (arteries) for a penetration of nanosize particles or their precursors into a near surface polymer layer. According to a size, pores are divided into three types: a micro pore ($r < 1.5$ nm), a mesic pore or a transient pore ($r = 1.5$ to 30 nm), and a macro pore ($r = 30$ to 6400 nm). Pores may be closed and transparent. As a rule, the polymer contains pores of various types, sizes, and forms. Fig. **5** shows a scheme demonstrating a nanoparticle formation in an implanted layer.

A method of Langmuir-Blodgett, which is an analog of molecular-beam epitaxy, allows a formation of two—dimensional, multilayered systems, and superlattices based on organic, biological molecules, and their combinations. Applying this technique, one can fabricate a nanosize organic and a bioorganic system on a hard substrate. For example, applying the above mentioned Langmuir-Blodgett method, a superfine film (~ 1 nm) was formed from a copolymer vinyliden fluoride with a trifluoride ethylene (PVDF, TFE), in which a phenomenon of two-dimensional ferroelectricity was revealed for the first time Fig. **6(a,b)**.

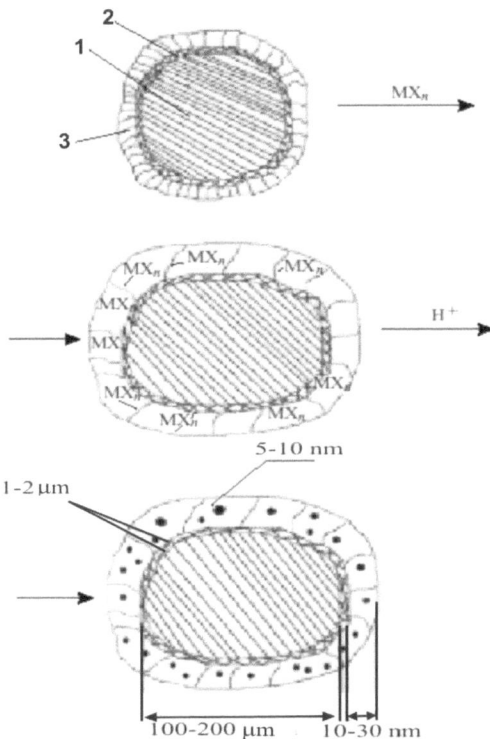

Figure 5: A common scheme demonstrating a formation of nanoparticles in an implanted layer: *1*-a polyethylene; *2*-a transient layer; *3*-an implanted polyethylene with an acrylic acid [13].

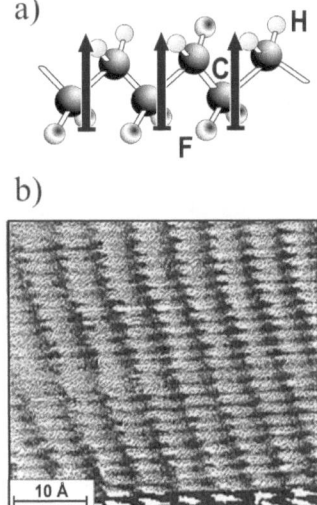

Figure 6: An illustration of ferroelectricity effect found in a superfine copolymer vinylidene fluoride with a trifluoride-ethylene (PVDF): *a*-a polymer molecule structure; *b*-a SEM microscopy image for a monolayer (PVDF, TFE) [16].

A synthesis of metal-containing polymers based on a low-temperature solid-phase polymerization of n-xylene monomers with a presence of various metals demonstrated that a monomer formed by a pyrolysis of corresponding n-cyclophane compound featured a high recreation activity in a solid state even at a low temperature. A conductivity of fabricated poly-n-xylene composite film with metallic nanoparticles can be essentially changed by an action of various chemical compounds of surrounding medium. Depending on a nature and a content of metallic nanoparticles, the film conductivity reacts with various compounds. Such films "work" as a selective and sensitive probe reflecting a state of surrounding medium. A nanoparticle of ferromagnetic material is a ferromagnetic monodomain in a material of polymer composite, which is isolated in a matrix of nonmagnetic polymer matter. Such films with a high content of nanoparticles open new promising ways for a fabrication of magnetic systems with a high density of both data recording and storage. For example the authors of works [16, 25-28] described an effect of magnetic field on a conductivity of fabricated film with partially oxidized iron nanoparticles containing an iron nucleus surrounded by a shell of iron oxides. Within a studied U interval (from 0 to 50 V), the conductivity at the interval boundaries changed very little due to an action of magnetic field and reached its maximum value 34% at about $U \sim 30$ V. When the magnetic field was switched off, the film conductivity quickly returned to its initial value, demonstrating that the magnetic field effect was reversible. An effect of negative magnetic resistance disappeared after a total oxidation of film nanoparticles,. This seems to be due to a specific character of electron transfer in a magnetic field between interacting nanoparticles having a magnetic momentum.

For example, properties of a two-phase system composed of metallic nanosize particles embedded in a non-conducting matrix will depend on a volume concentration of metal phase M. When $M > M_c \approx 0.5$, a percolation metallic cluster, which is a branched "network" of contacting with each other metallic particles, is stretched over the whole sample. The conductivity of such system has a metallic character. When a fraction of metallic phase is low ($M < M_c$), the conductivity is realized by a tunneling of charge carriers between nanocomposite individual particles (granules).

A nanocomposite material containing granules of a ferromagnetic metal have also interesting properties. When $M > M_c$, a sample is as an "infinite" ferromagnetic cluster, and its magnetic properties are similar to a three-dimensional metal with a definitely pronounced Curie temperature. A material is one-domain when granules have a small size (not more than ~ 10nm to 100nm, depending on a material). A direction of magnetic momentum of granules looks like a "game" between an orienting action of external magnetic field and a stabilizing action of magnetic anisotropy-a crystalline or a geometric one. A probability of tunneling transition between grains depends on a relative orientation of magnetic momentum of granules, which can be controlled by an external magnetic field. It results in a so-called effect of "gigantic" magnetoresistance, which is a very high relative change of nanocomposite resistance (in comparison with a usual metal) in a magnetic field reaching several scores of percent. Various properties of nanocomposite can be efficiently (*i.e.* significantly and reversibly) changed with the help of temperature or magnetic field. In principle, it opens new possibilities of a practical application. Thus, the authors of [22, 27-32], reported about a formation of new type nanocomposite material, which was a fine-dispersion metallic powder in a binding nonconducting material like an elastomer. The authors also pointed out that a conductivity of such material changed by several orders under various deformation types (a compressive strain, a torsion, and a tensile strain deformation). This phenomenon is associated with a physical nature of nanocomposite tunneling conductivity and seems to have an important future practical application. Materials with nanoparticles or metallic clusters distributed over a non-organic matrix have a special interest, which is associated not only with their application in a catalysis and an electrical engineering, but also with a development of "stellar-technologies".

QUESTIONS FOR CONTROL

1. What is a nanocomposite material?

2. What is a difference between a metallic and a polymer nanocomposite material?

3. What types of nanocomposite materials exist?

4. How can a polymer nanocomposite change its magnetic property?

5. Remember the ways, which are applied to form a metallic polymer composite.

REFERENCES

[1] Veprek S, Argon AS. Towards the Understanding of the Mechanical Properties of Super-and Ultrahard Nanocomposites. J Vac Sci Technol 2002; 20: 650-664.

[2] Niihara K, Nikahira A, Sekino T. Nanophase and Nanocomposite Materials. Mater. Res. Soc. Symp. Ed. By Komareneni S, Parker JC, Thomas GJ. Pittsburg. 1993; 286: 405-411.

[3] Ragulia AV, Skorokhod VV. Consolidated Nanostructured Materials; Kiev: Naukova Dumka 2007.

[4] Kelly P, Akelah F, Moet A. Reduction of Residual Stress in Montmorillionite Epoxy Compounds. J Mater Sci 1994; 28: 2274-2280.

[5] Chvalun SN. Polymer Nanocomposites. Priroda 2002; 7: 2-12.

[6] Kuntz JD, Zhan G-D, Mukherjee AK. Nanocrystalline-Matrix Ceramic Composites for Improved Fracture Toughness. MRS Bull 2004; 1: 22-27.

[7] Wilson JL, et al. Synthesis and Magnetic Properties of Polymer Nanocomposites with Embedded Iron Nanoparticles. J Appl Phys 2004; 95(3): 1439-1443.

[8] Dolgoshei VB. Thermal-Physical Properties of Nanostructured Polymers. Thesis Ph.D. Kiev: Institute of Chemistry of High-Molecular Compounds NAS of Ukraine 2002.

[9] Tuominen M, et al. Functional Nanostructures Based on Polymeric Templates. NSF Partnership in Nanotechnology Conference 2001.

[10] Strikanth H, et al. Magnetic Studies of Polymer-Shifted Fe Nanoparticles Synthesized by Microwave Plasma Polymerization. Appl Phys Lett 2001; 79: 45-51.

[11] Biswas A, et al. Controlled Generation of Ni Nanoparticles in the Capping Layer of Teflon AF by Vapor-Phase Tandem Evaporation. Nano Lett 2003; 3: 69-73.

[12] Gerasimov GN, Sochillin VA, Chvalun SN, et al. Cryochemical Synthesis and Structure of Metal-Containing Polyxylenes: System Poly-(Chloroxylene) Ag. Macromol Chem Phys 1996; 197: 1387-1393.

[13] Pomogailo AD. Metal-Polymer Nanocomposites with Controlled Molecular Architecture. Rus Khim Jour 2002; XLMI; 5: 64-73.

[14] Sergeev GB. Size Effects in Nanochemistry. Rus. Khim. Jour.2002; XLVI; 5: 22-29.

[15] Koval'chiuk MV. Organic Nanomaterials, Nanostructures, and Nanodiagnostics. Vestnik Rossiiskoi Academii Nauk 2003; 73: 405-409.

[16] Trakhtenberg LI, et al. Nanocomposite Metal-Polymer Films, Sensor, Catalyst, and Electro-Physical Properties. Vestn Mosk Uni Ser Khim 2001; 42: 325-331.

[17] Garsia M, Zhao Y-W. Magnetoresistance in Excess of 200% in Ballistic Ni Nanocomtacts at Room Temperature and 1000e. Phys Rev Lett 1999; 82(14): 2923-2926.

[18] Andrievskii RA. Nanomaterials: Concepts and Modern Problems. Rus Khim Jour 2002; XLVI: 50-56.

[19] Dementieva OV, et al. New Approach to Investigations of Surface Layers of Glass-Like Polymers. Butlerovskie Soobscheniia. 2001; 4: 1-5.

[20] Lagutin AS, Ozheghin VI. Strong Pulsed Magnetic Fields in Physical Experiment. Moscow: Energoatomizdat 1988.

[21] Tarasov KA, Isupov VP, Bokhonov BB, et al. Formation of Nanosized Metal Particles of Cobalt, Nickel, and Copper in the Matrix of Layered Double Hydroxide. J Mater Synth Process 2000; 8: 21-27.

[22] Decher G, Schlenoff JB. Multilayer Thin Films. Sequential Assembly of Nanocomposite Materials. Berlin: Wiley-VCH Verlag 2002.

[23] Koo J. Polimer Nanocomposites. Processing, characterization and Applications. USA: McGraw-Hill Companies Inc. 2006.

[24] Sung J, Lin J. Diamond Nanotechnology. Syntheses and Applications. Pan Stanford Publishing Pte. Ltd. 2010.

[25] Wong H-S, Akinwande D. Carbon Nanotube and Graphene Device Physics. Cambridge University Press 2011.

[26] Advani G. Suresh. Processing and properties of nanocomposites. World Scientific Publishing Co. Pte. Ltd. 2007.

[27] Birkholz M, Albers U, Jung T. Nanocomposite layers of ceramic oxides and metals prepared by reactive gas-flow sputtering. Surf Coat Technol 2004; 179: 279-285.

[28] Mai Y, Yu Z, Mai Y, Yu Z. Polymer Nanocomposites. Woodhead Publ. 2006.

[29] Pogrebnjak AD, Lozovan AA, Kirik GV. Structure and Properties of nanocomposite, hybrid and polymers coatings,Publ. House URSS, Moscow, 2011, 344

[30] Azarenkov NA, Beresnev VM, Pogrebnjak AD, et al. Fundamentals of Fabricated Nanostructured Coatings, Nanomaterials and Their Properties, Publ. House URSS, Moscow 2012; 352.

[31] Pogrebnjak AD, Sobol OV, Beresnev VM, *et al.* Phase Composition, Thermal Stability, Physical and Mechanical Properties of Superhard On Base Zr-Ti-Si-N Nanocomposite Coatings Nanostructured Materials and Nanotechnology IV: Ceramic Engineering and Science Proceedings 2010; 31(7): 127-138.

[32] Pogrebnjak AD., Uglov VV, Il'yashenko MV, *et al.* Nano-Microcomposite and Combined Coatings on Ti-Si-N/WC-Co-Cr/Steel and Ti-Si-N/(Cr3C2)75-(NiCr)25 Base: Their Structure and Properties Nanostructured Materials and Nanotechnology IV: Ceramic Engineering and Science Proceedings 2011; 31(7):115-126.

CHAPTER 6

Methods Employed for Nanomaterial Fabrication

Abstract: This Chapter considers the methods, which are applied for the fabrication of nanomaterials. They could be divided into 4 groups: powder metallurgy, controlled crystallization from an amorphous state, intensive plastic deformation, and thin film formation technology. Their classification is summarized in Tables. Methods employed for the fabrication of nanotubes are described. Probe systems applied for the formation of charged particle beams are considered. 32 Figures and 3 Tables are presented.

Keywords: Fabrication of nanomaterials, films, deposition, devices, application, classification.

6.1. INTRODUCTION

Technologically, methods, which are applied for a nanomaterialfabrication, are divided into four groups: a powder metallurgy, a controlled crystallization from an amorphous state, an intensive plastic deformation, and technologies for a thin filmfabrication.

Table **1** [1-3] presents methods applied for a nanostructure fabrication and nanostructure features.

Table 1: Basic methods applied for nanomaterial fabrication and structure features of these nanomaterials.

Technologies	Fabrication Methods	Materials	Structure Features
Powder Metallurgy	Gas-phase Deposition and Compaction. Compaction and Sintering. Hot Compaction, Forging, Extrusion	Metallic Materials, Ceramics, Composite Materials, Polymers	Porosity, Non-Equilibrium Interfaces
Controlled Crystallization from Amorphous State	Crystallization of Amorphous Alloys, Consolidation of Amorphous Powders with Subsequent Crystallization	Amorphous Metallic Materials	Subnanoporosity and Prismatic Dislocation Loops
Intensive Plastic Deformation	Equal Channel Angular Compaction. Torsion Strain under High Pressures. All-Round Forging. Phase Cold-Work-Hardening and Grain Grinding	Metals and Alloys	Internal Stresses. Non-Equilibrium Boundaries and Grain Junctions.
Fine-Film FabricationTechnology	Electrolytic Deposition. Chemical Deposition from Gas Phase. Physical Deposition from Gas Phase. Sol-Gel Technology	Metallic Materials, Ceramics, Composite Materials	2-D Dimension, Columnar Grains, Porosity

6.2. POWDER METALLURGY AND NANOMATERIAL FABRICATION

This technology may be conventionally subdivided into two groups: methods applied for a fabrication of nanopowders and for a production of compact tools from these powders. A number of methods may be used both for a fabrication of nanopowders and a production of bulk tools.

There is a number of general approaches, which characterize all methods applied for a nanopowder fabrication and make them different from those applied to a fabrication of a usual powder [2-6, 84-91]:

- A high formation rate of particle generating centers;

- A low rate of particle growth;

- The biggest size of resulting nanoparticles does not exceed 100nm;

- A narrow difference range of particle size, a stability in a formation of particles with a desired size;

- A reproducibility of chemical and phase particle composition;

- High requirements for a control and a management of fabrication process parameters.

Irrespective of an applied fabrication method, a characteristic feature of powder nanoparticles is their susceptibility to unite into an aggregate and an agglomerate. Therefore, not only an individual particle size but also that of the whole union should be taken into account. There is no a precise terminological difference between the aggregate and the agglomerate. However, it is known, that the aggregate intercrystallite bond is stronger and an intercrystalline porosity is lower. In comparison with a non-aggregated powder, a higher temperature and/or a higher pressure are necessary to reach a desired porosity of aggregated powder, which will be subjected to a subsequent compacting.

Let us consider all fundamental methods, which are applied now to a fabrication and compaction of a nanopowder.

An evaporation technology and a deposition from a vapor phase. Today, these methods are widely used to fabricate nanopowders. This is due to the fact, that a material evaporation technology using various high-intensity energy sources and a subsequent deposition from a vapor phase can be most easily controlled, in this way, they meet high requirements of a resulting powder purity, especially, of a chamber with a controlled atmosphere. Most often, the latter is filled in with an inert gas-a helium, an argon, or a xenon. When a metal is evaporated in a vacuum or the inert gas, a metallic atom, which transits to a gas (a vapor) phase, tends to unite into a particle of several nanometer size, which later is deposited to a cooled substrate. The above-mentioned group of methods allows the formation of a complicated alloyed powder. An alloys of desired composition can be formed both by an evaporation of preliminarily alloyed material and by a simultaneous evaporation of individual components. A particle size of resulting powder may range from 5 to 100 nm, depending on an applied method and technological parameters.

Depending on a type of evaporation process, one can identify the following groups of methods.

A thermal evaporation. An evaporated matter is heated in a melting pot. Today, there exist various ways of heating. As a rule, they all apply a high-intensity energy source: a high-frequency induction, an electron-beam, an electric-arc, a plasma source, and a laser source. Fig. **1** demonstrates a common scheme, which is applied to fabricate the nanopowder.

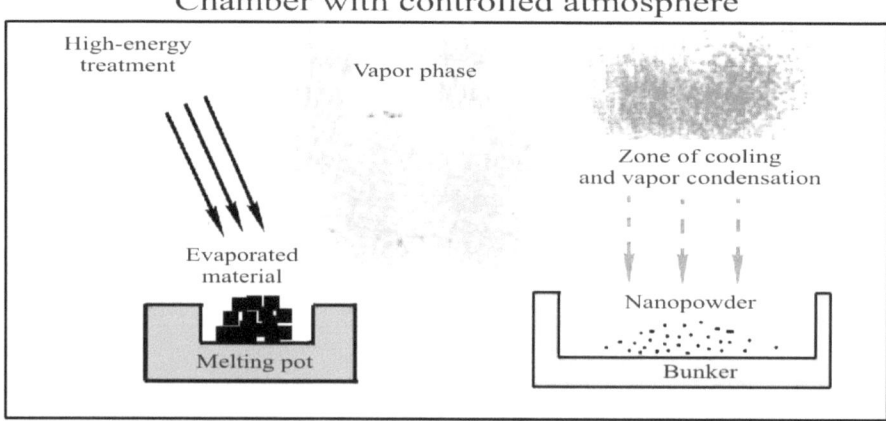

Figure 1: A common scheme of nanopowder fabrication by thermal evaporation and a material deposition from a vapor phase.

The resulting powder particle has a spherical or a cut form and may be a metal, an intermetalloid, or a compound. The electron beam thermal evaporation of bulky oxide in an inert atmosphere allows a formation of amorphous powder Al_2O_3, SiO_2, and crystalline Y_2O_3 [7-12].

An advantage of this method is a possibility to fabricate a pure powder with a low particle size difference, its disadvantage is a low process efficiency. This disadvantage seems to be temporal due to an absence of big devices for the nanopowder fabrication, which could be applied in an industry, rather than to a technology providing the fabrication process.

An explosion evaporation. Today, this method is quickly developing. It is based on an absorption of great energy amounts for a short time period. To generate a desired energy amount, a high-power impulse of electric current, an arc-discharge, or a laser pulse for material evaporation are employed. A treated material is evaporated as a result of such action and due to a quickly increasing bulk is cooled and vapors of a small-size particles are deposited. In a number of cases, some fraction of the material has no time for the evaporation. In this case, it is melted and explosively disintegrated into liquid drops. This method allows a fabrication of high purity powders with a size of spherical particles ranging from 5 to 10 nm, including metals with a very high melting temperature and a chemical activity. A disadvantage of this method is significant energy expenditures and a relatively expensive price of a powder, as well as difficulties in an elimination of micron-size particles coming from a melt. Fig. **2** presents a photo of NiO nanopowder fabrication by an electric explosion [13, 14].

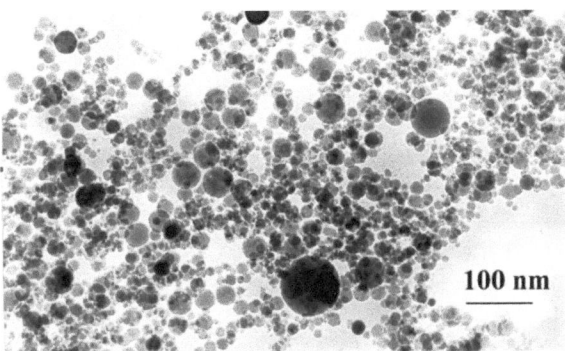

Figure 2: NiO nanopowder fabrication by an electric-explosion.

A levitation-jet method. This method is a metal evaporation in an inert gas flow. The metal can be evaporated, for example, from a drop of melt at a wire end heated by a high-frequency magnetic field. A device employed to a fabrication of the nanopowder using the evaporation in the inert gas flow is schematically shown in Fig. **3**.

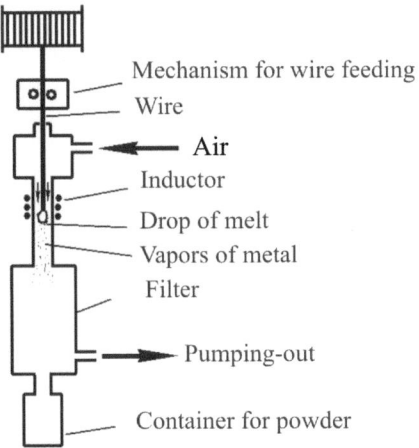

Figure 3: A scheme of device applied for a nanopowder formation using an evaporation in an inert gas flow [5, 7].

A size of resulting particles depends on a gas flow velocity-when the velocity increased, the size decreased from 500 to 10 nm, and, correspondingly, a difference in the particle size reduced. In particular, the above method is applied to a fabrication of Mn and Sb nanopowder. Due to a high quenching rate, the latter turned out to be amorphous. There is one more modification of the above method, which is called *a cryogen melting*. A wire is melted in a liquid of very low temperature, for example, in a liquid nitrogen [12, 15-17, 89, 93].

A plasma-chemical method. It is based on an application of low-temperature plasma induced by an arc or a glow discharge (an ordinary high-frequency or a super high-frequency discharge). A metal, halogen or other compounds can be used as an initial material. A transition of initial matter to a gaseous state is provided by a very high plasma temperature (reaching 10000 K) and interaction rate. Subsequently, they interact with each other, and a resulting product is condensed in a form of nanopowder. The nanopowder particle has a regular form and from 10 to 200 nm size. The highest temperature and power can be reached using an arc plasmatron. A nanopowder of highest purity and uniformity can be obtained using a super-high-frequency plasmatron. A nanopowder of carbide, nitride, boride, and oxides of various elements, as well as a many-component compound is fabricated using a plasma-chemical synthesis in an active media containing a carbon, a nitrogen, a boron, or an oxygen. A fullerene nanopowder can be obtained in the same way.

A chemical method. This technology is based on a chemical reaction of a metallic compound in a gaseous state. The compound starts a thermal decomposition in a certain zone of recreation chamber and forms a solid precipitate, which is the nanopowder or the gaseous matter, or starts a chemical reaction also resulting in a formation of powder or gaseous matter. A halogenides (mainly a chloride) of metals, an alkylated compound, a carbonile, an oxichloride may serve as an initial material. A size of resulting particle may be regulated by a temperature and a cooling rate. The same technology is applied to a fabrication of a silicon, a boron, a titanium, a zirconium, an aluminum oxide, a nitride, a carbide, a carbon nitride of silicon, and a titanium diboride nanopowder with 20 to 600 nm particle size [17-24, 85-90].

A sol-gel process was developed specially for the purposes of an oxide ceramics. This process includes the following stages: a preparation of alkoxide solution, a catalytic interaction with a subsequent hydrolysis, a condensation polymerization, and a further hydrolysis. As a result, an oxide polymer (gel) was formed. It was subjected to an ageing, a washing, a drying, and a thermal treatment. A disadvantage of this method is a complicated device assembling, and its advantage is a high purity and a uniformity of resulting synthesized compound and a possibility to form a great variety of nanopowders.

A compaction is a technological process allowing a fabrication of ready tools from a powder. Usually, this process is realized in two stages: a *compaction and a sintering*. In a number of cases, these both stages are united in one.

A compaction method. A uniaxial compaction technology is most widely used to compact a nanopowder. Such methods as: a static (a compaction in a mould or a stamping), a dynamical (magneto-pulsed and explosive), and a vibration (ultra-sonic) compaction. All the above mentioned ways of compaction are well known, are applied allover the world to compact a usual powder, and are well described in a literature [2-5, 11-13, 24, 89, 93].

A compaction with an all-round material compression is applied to obtain a high-density stamping. This technology was named an *isostatic compaction*. There are several ways to realize it: a hydrostatic, a gas-static, and a quasihydrostatic compaction. In the case of isostatic compaction, a powder is placed into an elastic or a deformable shell. A resulting mold features a practically uniform density (which sometimes is a little lower inside the mold volume) and does not have a pronounced anisotropy of its properties. A disadvantage of this technology is a significant complexity of equipment, a high cost, and difficulties in keeping an accuracy of the mold size.

A nanopowder mold sintering is limited by an impossibility of high temperature application. An increased temperature results in a decreased porosity, but on the other hand, it is able to increase a grain size. A number of activation methods, which can help to reach a low porosity of tools at a low sintering temperature, are employed to solve this problem:

- A high-rate microwave heating (a sintering temperature of TiO_2 nanopowder decreases from 1050°C to 975°C, when a heating rate increases from 10 to 300 grad/min);

- A step-controlled sintering;

- A plasma-activated sintering;

- A sintering in a vacuum or a reducing media (for a metallic powder).

A pressure sintering. A simultaneous application of molding and sintering (or a pressure sintering) yields a high density including a value, which is close to that deduced for a low heating temperature by theoretical calculations.

The simplest way is *a sintering under a uniaxial pressure.* For example, when an iron nanopowder is sintered under 400MPa molding pressure, a sintering temperature, which does not yield a porosity, decreased from 700°C to 350°C, and a tool grain sizes decreased from 1.2 µm to 80 nm. For a nanopowders with metallic particles, the process should be performed under vacuum conditions or in a reducing atmosphere.

A more productive way combining the molding and the sintering is *a hot isostatic compacting* (*HIC*). This method is universal and widely known in a practice of powder metallurgy. A modern device can reach up to 300MPa pressure and up to 2000°C temperature.

A directed compacting method is a cheaper modification, which can substitute the *HIC*. This method employs a heated thick-wall cylindrical press-form. First, this form is filled by a powder and then, is subjected to a uniaxial compression at a high pressure (up to 900 MPa). External walls of this press-form are adherent to a metallic cylinder corresponding to the press-form dimensions preventing in this way its deformation. The press-form inside volume, which has a form of future tool, is filled-in with a powder and subsequently is subjected to a quasistatic pressure. This method allows the compaction of metallic powder up to about 100 % density only for several minutes.

A high temperature gas extrusion method is a hydrostatic compaction at a room temperature, a subsequent treatment in a hydrogen medium under a relatively low temperature, and an extrusion at an elevated temperature [5, 24]. This method allows a powder compaction by a short-time action of essentially high temperature. For example, a compact, which is fabricated from a nanopowder based on a nickel particle using this method, is characterized by a high strength and a very good plasticity index, simultaneously.

6.3. AMORPHOUS MATERIAL FABRICATION

An amorphous metallic alloy (*AMA*) is a new promising class of materials. The amorphous alloys is characterized by an absence of long-range order in an arrangement of packing atoms. Such state is reached under conditions of a superfast cooling from a gaseous, a liquid, or an ionized material state [21-23].

The following methods are applied for the amorphous alloy fabrication:

A quenching from a liquid state. One of the most common ways to a fabrication of *AMA* is a cooling with about 10^5 to 10^8 K/s rate from a liquid material state. Therefore, a characteristic feature of this method is a fast melt cooling without a crystallization. In practice, the crystallization can be prevented and a glass-like state can be fixed if a liquid melt contacts to a cool metal substrate. Two methods are most popular in practice: one of them is a deposition of liquid metal to an external cylindrical surface of rotating disk (a wheel), the second one is a melt extraction by a rotating disk. Fig. **4** shows a scheme of these two methods. A rim of the metallic disks or cylinders should be made of a material with a good conductivity.

Usually, a copper, a beryllium bronze, a brass, *etc.* are applied for this purpose. The melt is heated by an induction heating device or by a resistance furnace. An induction nozzle is made of a melted quartz or an aluminum oxide material.

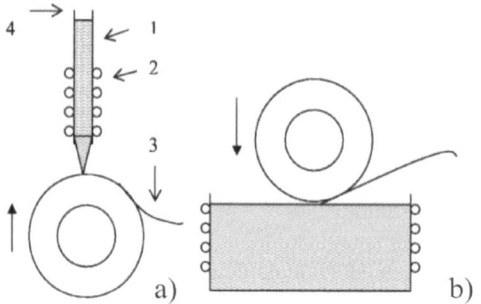

Figure 4: Schemes of *AMA* fabrication using a quenching from a liquid state: *1*-a melt; *2*-a heating unit; *3*-an *AMA* ribbon; *4*-a quartz tube.

A foil, which is fabricated using this method (Fig. **4a**), looks like a continuous ribbon of 1nm to 200mm width and 20 µm to 60 µm thickness. The second method involves that a disk rim captures a melt, which is later solidified and spontaneously separated (Fig. **4b**). A thin wire, a cross-section profile of which depends on the disk pointed edge profile and its emersion depth into a melt, is formed.

The fabrication of the amorphous alloys ribbon by quenching from a liquid state at room temperature and under conventional pressure of surrounding atmosphere, should meet certain main requirements:

1. A volume velocity of melt V_T flowing through a hole in the rotating disk surface must be constant over a whole period of ribbon formation.

2. A melted jet flow must be stable and defended from an action of small dust particles and a non-controlled air flow.

3. A resulting surface must be polished and have a good thermal and a mechanical contact with the melted jet.

In addition, a resulting quality and a width of amorphous ribbon is greatly influenced by a gaseous medium, in which they are formed. A relatively narrow ribbon may be fabricated under conventional pressure of the surrounding gaseous medium. Often, a wider ribbon formed under such conditions has a jagged edge, a non-uniform thickness, roughness, and through holes. In practice, a boundary layer of the rotating disk surface plays a great role in a quality of fabricated *AMA* ribbon. Due to a friction power, a velocity of gas molecules moving near the disk surface may be equal to a disk velocity. Therefore, such gas flow (or a liquid one) characteristic as a Reynolds number R_e plays a great role in a resulting film quality. A smooth ribbon edge and surface is formed when the Reynolds number of gaseous boundary layer is smaller than some critical value $R_{ek} < 2000$. At $R_e > 2000$, a kinetic energy is scattered due to an arising turbulence resulting in the formation of jagged edge and non-smooth surface. Indeed, the turbulence may be fully avoided if a ribbon is fabricated under vacuum conditions, when a residual pressure does not exceed 10^{-2} Pa. Perhaps in this case, conditions for the fast quenching will be worse, however, the *AMA* fabrication under vacuum conditions seems to be more preferable technologically, since it allows a better quality of fabricated materials, which can undoubtedly find a wide industrial application. A melting pot and a nozzle for a flowing-out melt is also an important unit playing a definite role in the amorphous ribbon fabricating. Usually, a quartz or an aluminum oxide is used for their fabrication. A diameter of the nozzle hole may vary within 0.4 to 2 mm. An open-end of the nozzle is closely positioned to the rotating disk surface. In general, one should note that the closer was the dram or the disk surface from the nozzle opening-end, the lower was the flowing-out jet turbulence.

Let us say several words about a surface treatment applied for a cooling disk. A copper disk surface has a low hardness, and prior to the ribbon fabrication, it must be polished and purified. With the purpose to increase the surface hardness, the disk surfaces is coated by a chromium. Experiments demonstrate that a diameter of the cooling disk or a dram should be about 200 nm or more. Such diameter yields a high linear rate under conditions of slow rotation and decreases a disk vibration during its operation.

Let us consider briefly several additional ways applied for the *AMA* fabrication.

A deposition to a cooled substrate using an ion-plasma and a thermal sputtering. This deposition method allows a fabrication of amorphous structure with a complicated composition of about 1cm thickness on a substrate quenched up to a temperature of liquid nitrogen. This method allows a formation of materials with various compositions, since both alloys prepared by ordinary metallurgic methods and specially prepared targets can be sputtered. Also, we should like to note that the ion-plasma deposition allows a fabrication of such bulky amorphous state, which cannot be fabricated by the quenching from a liquid state.

A fabrication of amorphous metallic layer by a laser treatment. An amorphous structure is formed by a short-time interaction of high energy density laser radiation with a matter. A main fraction of this energy ($\sim 10^3$ to 10^{10} W/cm^2, depending on a material) is spent to a fast melting of a material surface layer. Because this process is momentary, the main material mass remains non-heated. A layer, which contacted to the cool metal surface, quenched with 10^5 to 10^8 K/s rate. A highly homogeneous liquid, which was formed as a result of "fast" melting, after a solidification was transformed into a "glass" with an unusual physical property. This process is called "a laser-induced glassing".

An electric field sputtering. A strong electric field generated between a liquid melt and a surface can induce a sharp protuberance emitting both an ion flux from the melt and drops of 0.1 μm to 20 μm size, which are quickly cooled and form the *AMA*. This method can be applied to a fabrication of an amorphous coating and powder.

An ion implantation. An amorphous structure can be fabricated by a doping of high-energy ions into a surface layer of metallic alloy. This method is good to a fabrication of corrosion resistant and hardened surface. A main disadvantage is a low thickness of resulting amorphous layer, which usually does not exceed 1 to 2 μm.

An electric-arc discharge amorphization. An essence of this method is that an energy flow, which is concentrated in a channel of spark discharge, is released for a short time period ($\sim 10^{-3}$ s) and melts some material surface region to 3-5 μm depth.

It is known, that an alloy in an amorphous state is metastable. Therefore, after the amorphization, it needs an annealing, in the process of which it partially transfers to a more stable state. However, the metastability still remains, and in the process of heating to (0.4 to 0.65)T_{melt}, a material transits to a crystalline state. In this connection, not only the amorphous material state but also a nanocrystalline state of alloys started to attract a great interest.

A nanocrystalline or amorphous-nanocrystalline structure of a number of bulky-amorphous alloys being capable of bulk-amorphization (for example those based on an iron) can be formed directly in the process of melt quenching when a quenching rate is a little lower than a critical rate of amorphous state formation. However, the majority of alloys demonstrate a non-uniform and non-stable structure. A formation of nanocrystalline structure as a result of crystallization occurring in the process of amorphous material deformation seems to be very promising.

6.4. METHODS OF INTENSIVE PLASTIC DEFORMATION

This group of methods for a nanostructured material formation is based on a plastic deformation of a great deformation power under a high pressure and at a relatively low temperature. The above-mentioned deformation conditions result in a strong grinding of a metal and an alloy microstructure reaching a nanosize range. These methods imply a number of requirements: a dominating formation of a structure with ultra small grains and big-angle grain interfaces (in this case, one can observe a qualitative change of material properties), a stability of material properties, which can be provided by a uniform nanostructure distribution over a total material volume, an absence of mechanical defects and cracks independent on intensive plastic deformation of material. This group of methods allows a fabrication of a volume, pore-free metallic nanomaterial. However, one should note that as a rule, a grain size of materials, which are fabricated by the above-considered methods does not exceed 100 nm. A structure formed by the intensive plastic deformation, is highly non-equilibrium due to a low density of free dislocations and a dominating big-angle character of grain interfaces. Therefore, a treated tool should be additionally subjected to a thermal treatment or the plastic deformation at a higher temperature and by a high deformation power. Today, the following methods are most popular.

A twisting under a high pressure. This method applies a Bridgeman anvil. A sample is placed between blocking heads and compressed under a pressure of several GPa (Fig. **5a**). A lower blocking head is rotated, and a forces of surface friction induces a shear deformation of a sample. Geometrically, the sample is a disk of 10nm to 20 mm diameter and 0.2 nm to 0.5 mm thickness, which condition a nondestructive hydrostatic compression of a basic material volume. The material structure starts to be grinded even after a half-revolution deformation. A structure with ultra small grains is formed in several revolutions. An average grain size may reach 100 to 200 nm depending on the deformation conditions: a pressure, a temperature, a deformation rate and a type of treated material.

An equal-channel angular compaction (Fig. **5b**) method allows a fabrication of big tools with 60mm diameter and 200 mm length (Fig. **6**). This method is also based on the shear deformation. A billet in a special rigging is forced many times through two crossing channels of equal cross-sections at room temperature or a little higher, depending on a treated material.

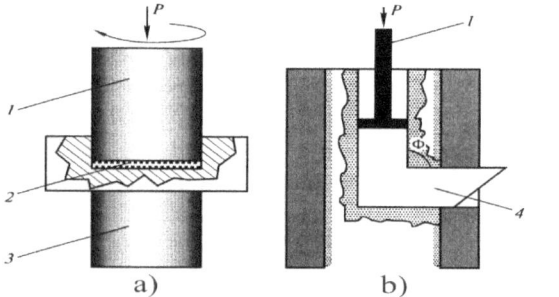

Figure 5: Methods of intensive plastic deformation: *a*-a twisting under a high pressure; *b*-an equal-channel angular compaction [6, 24].

An important problem is to keep a resulting sample integrity in the case when the sample has a low plasticity and difficult deformation response. The method allows a formation of a structure with an ultra small grain of 200 nm to 500 nm size [6].

Other methods of intensive plastic deformation are under development, for example, an all-round forging and a special rolling.

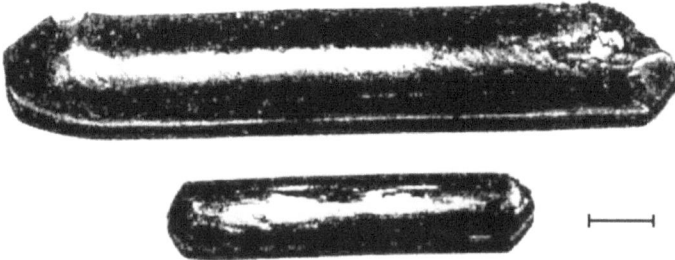

Figure 6: Bulk billets of a nanostructured titanium [6].

6.5. THIN FILM TECHNOLOGIES FOR SURFACE MODIFICATION

Today, a material surface treatment is one of the most intensively developing fields of material science. Methods relating to a modified layer fabrication in a material surface (especially a metallic one) are well known, developed, and widely applied in a practice [53-56]. Many of these methods or their advanced versions may be considered as a nanotechnology, since they allow a fabrication of a nanosized and/or nanostructured layer in a material surface, a composition material with a nanocomponent, and, in a number of cases, a nanomaterial working like a nano-or microtool.

These methods may be conventionally subdivided into two big groups-a technology based on a physical process and that based on a chemical process. Today, an ion-vacuum coating deposition technology (*PVD* and *CVD*) is most promising one among all nanooriented technologies applied for a surface treatment [37-40]. A layer fabricated by this technology is characterized by a high adhesion and a minimum temperature response. An analysis of literature data reported in [40] demonstrated that a crystallite size of a film, which was fabricated using the vacuum deposition technology, might reach 1 to 3 nm.

Let us consider in details some basic nano-oriented technologies, which are applied for the surface treatment.

6.5.1. Physical Deposition from Vapor Phase (PVD)

This group of methods is often designated as a *PVD* (a Physical Vapor Deposition) [25-37]. This group of methods has a common scheme of the coating deposition and a mandatory vacuum condition. First, a material, which is intended to be a coating, is transformed from a condensed-to-vapor state. Then, a vapor is transported to a substrate (a material, which is intended to be coated), covers it, and forms a coating. The vacuum condition makes easier a material transition to a vapor phase.

The PVD method is employed to form a very uniform surface layer of < 1 nm to 200 μm, with a very good reproducibility of its properties, to coat a surface (a magnetron method) with a practically unlimited length, to realize a selective deposition to an individual region, to fabricate multi-layered coatings, layers of which have a different thickness and are composed of different materials, to form a composition, a structure, and other layer properties using a variation of deposition technological parameters, and to avoid a contamination of surrounding medium. A disadvantage of these methods is a complexity and a high cost of applied technological and controlling equipment, a very high requirements to a staff qualification, a comparatively low efficiency, a problem in a development of technological regimes for an every concrete case of coating deposition, especially, if a coating is fabricated from a compound with a simultaneous high accuracy of its chemical composition, and a special surface treatment.

A thermal evaporation. This method is based on an initial material heating with the help of an energy source (a resistive heating, a heating by an electron beam or an electric discharge, *etc.*). The material is heated to an evaporation temperature, or it may be evaporated with a subsequent condensation, when its vapors form a thin film or a coating on a solid surface [35]. The matter transits to a vapor phase as a result of evaporation or sublimation. The vapor, which is in a thermodynamic equilibrium state with a liquid or a solid, is called saturated. The evaporation occurs when a kinetic energy of molecules and atoms of a solid or liquid surface layer at a given temperature exceeds their average volume energy so much that they tear off and spread in a free space. The atomic energy depends on an evaporator temperature and reaches 0.1 to 0.3 eV. A driving force of a particle transfer is a pressure difference between a saturated vapor above an evaporation surface and near a condensation one. An evaporation is intensified when a heating temperature increases. A dependence of saturated vapor pressure on a temperature for a one-component system is expressed by a Clapeyron-Clausius equation:

$$dP/dT = \Delta H_{ev}/T(V_v - V_l), \qquad (1)$$

where dP is a change of vapor equilibrium pressure due to a little temperature change d_T; ΔH_{ev} is a matter evaporation heat; V_v and V_l are a molar volume for a vapor and a liquid, respectively.

Assuming that $V_v \gg V_l$ and a vapor follows a law of an ideal gas, i.e. $V_v P = RT$, an equation (1) after a corresponding transformation looks like:

$$\lg P = -\Delta H_{ev}/RT + \text{const}, \qquad (2)$$

where R is a universal gas constant.

The equation *(2)* is valid for a narrow temperature interval, since it does not take into account a dependence of evaporation heat on a temperature. Physical reference books cite an equation allowing one a precise pressure value for a broader temperature interval taking into account a change in thermal-physical characteristics:

$$\lg P = A/T + B\lg T + CT + D. \qquad (3)$$

However, various metals have various pressure values of saturated vapors at the same temperature limiting, in this way, an application of thermal evaporation method for a formation of alloy coatings. Technological procedures allowing a fabrication of coatings with a stoichiometric composition were developed. They were the following: an initial composition could be changed with the purpose to compensate a difference in vapor elasticity, a measured amount of matter amount could be evaporated, an explosion-like evaporation or an evaporation from several melting pots could be employed. The vapor elasticity does not practically depend on a surrounding gas pressure. The pressure specifies a vapor diffusion from a boundary layer above an evaporator and, correspondingly, an evaporation rate. When the pressure in a chamber is low ($P \leq 10^{-2}$ Pa) and an average molecule free range does not exceed a characteristic scale, an effect of residual gas may be neglected, and, according to a gas kinetic theory and a Hertz-Knudsen equation, with respect to a mass, the evaporation rate will be determined by a Langmuir equation:

$$W = 7.78(M/T)1/2P, \qquad (4)$$

where W is an evaporation rate, $gcm^{-2}s^{-1}$; M is a matter molar mass, g/mol; T is an evaporation temperature, K; P is a pressure of a saturated vapor.

The pressure of a residual gas in a working area essentially affects a character of atomic flow distribution. A collision with a gas molecule changes the atomic flow initial energy and its trajectory. In the case of molecular flow excluding the collision of molecules themselves and atoms, a Lambert-Knudsen law describes a distribution of evaporated atoms. According to the first law, a vapor flow intensity towards φ direction, which is an angle between a normal to an evaporation surface and a direction of evaporated atoms, is proportional to a cosine of this angle. The evaporation is directed along a normal to the evaporation surface, i.e. a highest $\cos(\varphi)$ value. A non-uniformity of vapor flow distribution, in its turn, results in a non-uniformity of a coating thickness. An amount of deposited matter is inversely proportional to a square of a distance from an evaporator to a deposited surface (the second law). In practice, to obtain a deposition zone of a larger area with a uniform coating thickness, the evaporator-substrate distance should be scaled up, the evaporator surface should be enlarged, and a special inside chamber equipment, which could provide a motion and a rotation of tools in the process of deposition, should be developed [33]. A material evaporation temperature specifies a technique, which is employed for its evaporation. The material can be heated resistively, by a high-frequency electromagnetic field, by an electron bombardment, a laser emission, or an electric discharge.

The majority of these methods are employed for an evaporation of metallic materials. The heating in a melting pot and an application of laser emission are employed to evaporate a wide spectrum of materials. Thus, the heating in a melting pot is applied to evaporate a material with a relatively low evaporation temperature, which is specified by a temperature and chemical resistance of the pot material. So, a graphite melting pot can endure 1400°C, Al_2O_3 one to 1600°C, that of BN +TiB_2 to 1750°C, Mo or Ta foiled one, with a protective coating to 1850°C, and a pot of ThO_2 or ZrO_2 to 2100°C [33]. An important requirement for the melting pot material is an absence of any chemical interaction between the melting pot and the matter, which is evaporated at a high temperature [37-40].

The electron beam evaporation [41-43] is employed to improve evaporation conditions and overcome a number of other disadvantages of the melting pot evaporation. In this case, an electroconductive material is evaporated in the melting pot with a water-cooling. Then, it is heated by an electron beam of 2kV to 10 kV beam accelerating voltage and about 0.1A current. A disadvantage of the electron beam and the melting pot evaporation is a complexity, because the material components have a different vapor elasticity at the same temperature causing a problem with a fabrication of coatings of a desired chemical composition.

A laser emission (pulsed or continuous) is free from the majority of temperature and chemical limitations and does not employ the melting pot. A ratio of chemical components in a deposited film after a prompt matter evaporation remains practically the same as it was in an initial material. Not long ago, an application of laser emission for these purposes was restrained by a high cost of a high-power pulsed or continuous beam laser and an adjustment complexity of an optical system applied for a transportation, a focusing, and a guidance of the laser beam.

In a number of cases, a working temperature, which is necessary for an intensive evaporation and a desired process efficiency, is lower than a melting temperature of evaporated material due to vacuum application. Usually, a temperature, which can provide not lower than 1 Pa (10^{-2} mm Hg) of a steady-state pressure of material vapor, is employed to evaluate a working temperature of heating [35]. For the majority of materials, a working evaporation temperature ranges within 1100°C to 2600°C. A coating deposition rate may reach from several angstroms to several microns per second (for example, it is up to 5 μ/s for W and up to 40 μm/s for Al) [30, 35]. In a number of cases, a substrate heating is employed to improve an adhesion or form a definite structure of a deposited coating. As a rule, every individual component is evaporated from an individual source when a coating is formed from an alloy or a compound, since a vapor pressure value of every component of a complex matter may be vigorously different. In this case, a vapor phase and consequently, a coating composition will differ from that of evaporated matter. In addition, a compound evaporation is often accompanied by a dissociation and/or an association process, which also prevents the formation of a desired coating composition. The evaporation is applied only in the case, when a volatility of matter components is equal, and when a matter transits to a vapor phase in the form of non-decomposed molecules.

An advantage of the thermal evaporation method is a relative simplicity of equipment and a possibility of process control. Its disadvantage is a low coating adhesion due to a low energy of atoms or molecules deposited to a substrate and a high sensitivity to foreign films and contaminations occurring in the substrate surface. These disadvantages may be eliminated by special methods of the surface preparation (an ultrasonic surface purification, a chemical or an electric-chemical purification, and/or an etching, and an ion etching).

The thermal evaporation is extensively applied for a fabrication of computer hard magnetic disks [5]. An aluminum disk with not less than 20 nm height of surface micro-irregularities, which is coated by an amorphous nickel-phosphorous sublayer of about 20 μm thickness (to improve an adhesion and to compensate a difference in a thermal expansion coefficient of a substrate and a coating), serves as a substrate. First, a middle metallic layer (for example, Ni-Fe of 500 nm to 1000 nm), then, a basic layer of magnetic material (for example, an alloy based on Co or Co-Cr of 100 nm to 500 nm thickness), and, finally, a wear resistant carbon layer of 30 to 50 nm thickness is deposited.

The thermal deposition is applied for an industrial production of CD disks [5]. A plastic disk is coated by aluminum of ~ 300 nm to 500 nm thickness. In both cases, not less than 10^{-5} Pa pressure should be reached in a vacuum chamber to provide a high material purity. The considered method is also applied in an electron-optical engineering, which needs a regular nanostructure including two-dimensional photon crystals (like a fullerene) or composite fullerene-based films [44, 45].

Recent years, active practical studies are concentrated at a thin coating and a layered composite material fabrication using a very short laser beam (up to a femtosecond range). In literature, this method is designated as a *PLD* (a pulsed laser deposition). For example, Y_2O_3-ZrO_2 film on a silicon base [46], a layered composite (composed of Sm-Fe coating (20 nm), Ta sublayer (100 nm), and a silicon substrate) [46], a magnetic Ni film with 40nm average crystallite dimension [47] are fabricated in this way.

A cathode sputtering. A general scheme of device for a cathode sputtering is presented in Fig. 7.

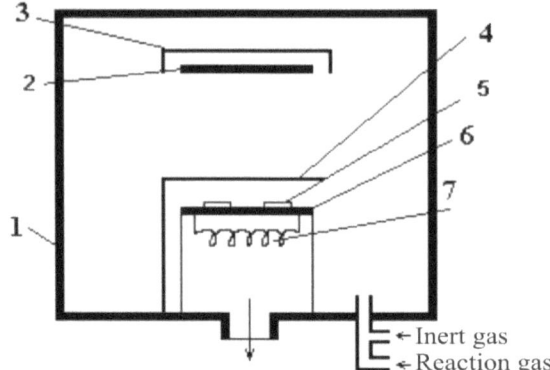

Figure 7: A scheme of a device for a coating deposition employing a cathode sputtering: *1*-a chamber; *2*-a cathode; *3*-a grounded screen; *4*-a gate; *5*-a substrate; *6*-a grounded anode; *7*-a resistive substrate heater [48].

This method is realized in the following way. A vacuum volume, in which an anode and a cathode are located, is pumped out up to 10^{-4} Pa pressure. Then, an inert gas is fed inside (usually Ar of 10^{-1} Pa). To burn a glow discharge between the cathode and the anode, a high voltage of 1 to 10 kV is applied. Positive ions of the inert gas, which come from the glow discharge plasma, are accelerated in an electric field, bombard the cathode, and induce its sputtering. Sputtered atoms come to a substrate, cover it, and form a thin film. The cathode sputtering is mainly used for a fabrication of metallic material layers.

A magnetron sputtering. A device employed for a magnetron sputtering is presented in Fig. 8.

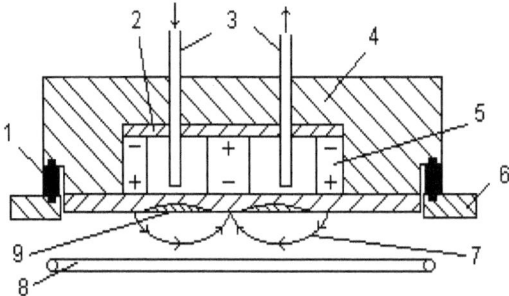

Figure 8: A scheme of a magnetron system employed for ion sputtering with a plane cathode: *1*-an isolator; *2*-a magnetic core (a circuit); *3*-a water cooling system; *4*-a cathode site body; *5*-a constant magnet; *6*-a vacuum chamber wall; *7*-a magnetic field line; *8*-a ring anode with a water cooling; *9*-an erosion zone of a puttered cathode.

A flat cathode, which is fabricated of a material intended for a future deposition, an anode, which is arranged over the cathode perimeter, a magnetic system composed of constant magnets, and a system for a water-cooling are basic elements of this device. A magnetic field line is closed between the poles and crosses an electric field line. A principle of device functioning is based on an electron deceleration in electrical and magnetic crossed fields. It is known, that a Lorentz force affects a charge moving in an electromagnetic field. A direction of this force, according to a force sum rule, depends on a direction of its components, but a fraction of it is affected by the electromagnetic field and does not work, bending only a particle motion trajectory and making a particle move along a circle inside a plane perpendicularly to E and

B. In such a way, the magnetron device changes the electron motion trajectory due to a simultaneous action of the electrical and the magnetic field. Electrons, which are emitted from the cathode, and those, which are formed as a result of ionization, are localized immediately above a sputtered material surface due to an action of closed magnetic field. They as if get into a trap. On one hand, the magnetic field makes the electron move along a cyclic near-surface trajectory. On the other hand, the cathode electric field repels them towards the anode. A probability and an amount of electron collisions to argon molecules as well as their ionization sharply increase. An ionization of various near-cathode regions is different since the plasma is non-uniform. The ionization maximum value was observed in a region where the magnetic field induction line was perpendicular to the electric field intensity vector. The ionization minimum was in a region where both directions coincided. A high ion current density at lower working pressure and, consequently, a high sputtering rate were reached due to the plasma localization in the near-cathode region.

The magnetron device is a low-voltage system employed for an ion sputtering. A direct current source voltage does not exceed 1000 V to 1500 V. When a negative potential is switched to the cathode, an anomalous glow discharge is excited between the electrodes in an argon medium. A magnetic trap allows a formation of discharge at a similar gas pressure and at a lower voltage in comparison with a diode system. The discharge voltage can reach 300 to 700 V. The magnetron device functions at 10^{-2} to 1 Pa working gas pressures and higher. The gas pressure and a magnetic field induction are important for discharge characteristics. A decreased pressure increases the working voltage. At the same time, every magnetron system has a certain pressure interval, usually 10^{-1} to 1 Pa, which does not affect essentially an alternation of the discharge parameters. A magnetic field action is similar to that of a gas medium. Therefore, a low working pressure increases a magnetic field induction. Its value near the cathode surface is 0.03 to 0.1 T. An increased specific power stabilizes the discharge in a low-pressure range [35, 48].

Advantages of this method are the following [49, 50]:

- A high sputtering rate at a low working voltage (600 to 800 V) and a non-high working gas pressure (5×10^{-1} to 10 Pa);

- An absence of substrate overheating;

- A low degree of film contamination;

- A possibility to fabricate a film with a uniform thickness stretching over a large substrate area.

Fig. **9** presents an electron microscopy image of a surface relief and a transverse chip of ZnO sample fabricated by the magnetron sputtering. Fig. **9a** shows the close-packed nanocrystalline ZnO layer with a relatively non-high surface roughness (Fig. **9a**) formed in parallel to a substrate plane (Fig. **9b**). A photograph of the surface micro-relief and the cross-section chip and an X-ray analysis of the studied sample confirm a high optical quality of the crystalline ZnO film and reproducibility of its structure, irrespective of a conductivity value.

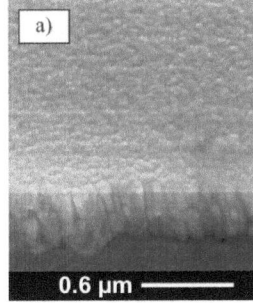 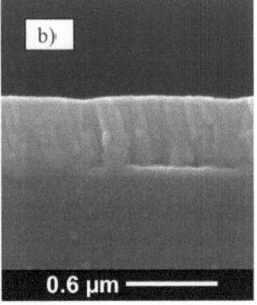

Figure 9: A photograph of ZnO-film micro-relief (x50000) formed by a magnetron sputtering: a)-a vertical cross-section; b)-a surface [5].

A vacuum-arc deposition. A vacuum-arc deposition employs a generation of highly ionized metallic plasma flow from an evaporated material using a vacuum arc. A high negative potential is applied to a substrate and provides an efficient purification, an activation, and an atomic diffusion. The coating adhesion to the substrate is high, in comparison with the magnetron method. A reaction gas fed to a vacuum chamber allows a fabrication of coatings based on a compound featuring high physical and mechanical properties. A substantial difference between the vacuum-arc and the magnetron method is that a plasma flow contains drops of evaporated material, which affect a resulting coating structure and generate additional distortions, interfaces, and pores. A total number of plasma filters was designed to decrease the drop fraction. One more step in the development of vacuum-arc technology is a coating deposition using a plasma ion implantation [51]. This method is realized using the following typical electrical scheme: a constant and single-polarity negative potential of an alternating frequency and amplitude (Fig. 10) is applied to a substrate.

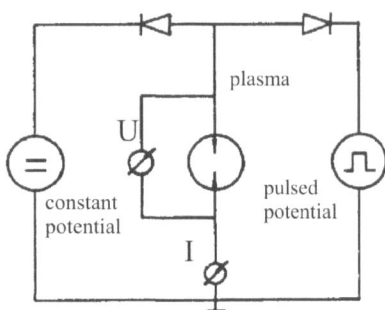

Figure 10: A scheme of switching a constant negative and a pulsed potential.

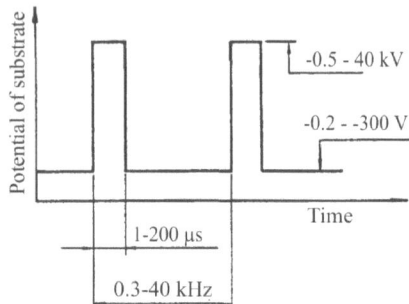

Figure 11: A time dependence of a superposition of a constant and a pulsed potential.

Fig. 11 demonstrates a typical time dependence of a combined potential cited in the work [52]. The potential of treated object is a superposition of a constant negative and a negative pulsed potential.

The first and now most important achievement of the new technology is that a temperature of TiN coating synthesis was substantially decreased to 100°C-150°C. Due to this fact, a titanium nitride coating can be now deposited to all types of structural and tool steels, including those, which have a low tempering temperature.

A method "Plasma-based ion implantation and deposition" *(PBII&D)* can provide the highest adhesion among all known *PVD* deposition methods. This highest adhesion was reached due to a thin transition layer (not an intermediate one, as it was earlier), which was formed between a substrate and a surface layer [53]. This new method allows an efficient control of a a value of compressing stress arising in a coating in the process of low temperature synthesis. The ion implantation of relatively low energy (0.5keV to 5keV), may be efficiently applied to decrease an internal stress, which depends on a product of the pulse amplitude and the repetition frequency. An internal stress of TiN coating can be decreased to 1GPa by decreasing the value of this product. TiN coatings of 21GPa microhardness and 0.9GPa to 2.9GPa internal stress were formed using a constant negative bias of 75V and a negative pulse of 5kV amplitude, 1μm to 3 μs duration, and 1kHz to 2kHz repetition frequency.

A hardness is an important characteristic of a tool wear resistance. An analysis of processes induced by an ion implantation in a deposited coating indicates that a superhard TiN coating can be formed at a substrate temperatures of about 100°C.

A technological complex [54] based on the vacuum-arc method (Fig. 12) was constructed for the coating deposition by the ion bombardment.

A molecular gas first passed through a cylindrical quartz discharge chamber for an additional chemical activation. A periodically repeated spark discharge was generated in this chamber by the HF generator. An arising shock wave compressed and heated a gas along the discharge chamber axis resulting in a molecular gas dissociation. An atomic gas adiabatically expanded in a technological volume without a recombination after this molecular gas dissociation. A tool (8) was arranged on a movable substrate, and the applied to substrate HF voltage was fed through a matching device (5, 6) from the HF generator (7). A new technological scheme allowing a fabrication of coatings from a metallic plasma flow was developed. A damping HF oscillation created a condition for the ion bombardment (an implantation) at the beginning of pulse and for a subsequent ion deposition in the surface. In such a way, the coating is deposited for one pulse independently on the device working characteristics (a partial pressure of a working gas, working conditions of a plasma source, *etc.*). To amend technological system, the simplest generator allowing one to reach 100kWt HF power (under 10kWt average HF power) per pulse was developed. It was based on generator with shock excitation [55].

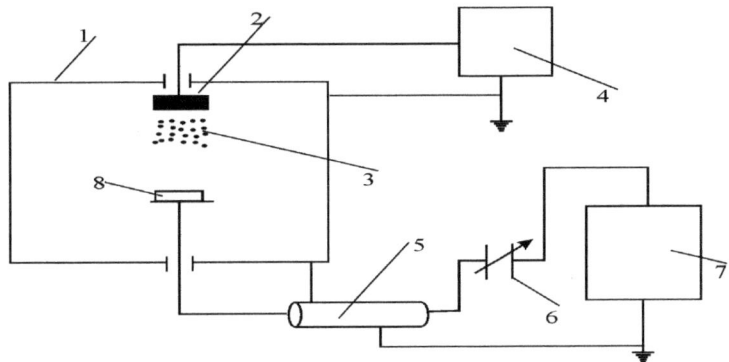

Figure 12: A scheme of technological system based on a vacuum-arc discharge employed for a coating synthesis: *1*-a vacuum chamber; *2*-an evaporated material; *3*-a plasma flow; *4*-a source feeding arc-evaporator; *5*-a coaxial cable; *6*-a variable capacitor; *7*-an HF generator; *8*-a tool.

A generator with a pulsed voltage, a regulated impulse amplitude and duration, and a controlled repetition frequency, as well as an electron control system with a regulated layer periodicity were developed for the purposes to fabricated multi-layered nanostructured coatings. A scheme of this modified device is presented in Fig. **13** [51].

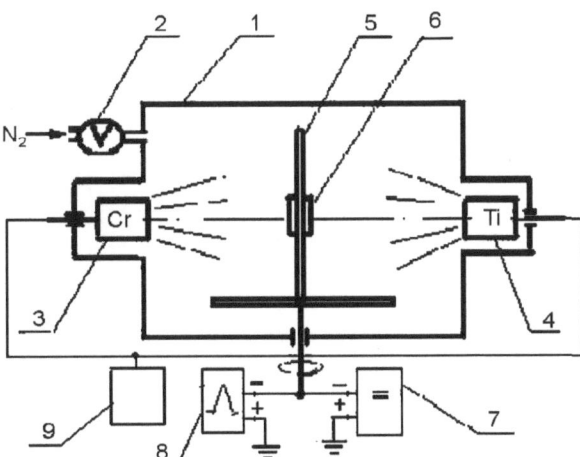

Figure 13: A scheme of a device for a deposition of a multi-layer two-phase nanostructured TiN-CrN coating: *1*-a vacuum chamber; *2*-a system of an automatic control for a nitrogen pressure; *3*-a chromium evaporator; *4*-a titanium evaporator; *5*-a substrate holder; *6*-a substrate; *7*-a constant voltage source; *8*-a pulse generator; *9*-a computing device.

An ion-beam sputtering, in fact, is an advanced version of a cathode and a magnetron sputtering. A difference between them is that ions of an inert gas are fed to a sputtered material (a target) from a separately arranged independent ion source in the form of a concentrated flow of 1keV to 10keV energy [56-58], Fig. **14**. A process runs in a vacuum of 10^{-3}Pa to 10^{-2} Pa, since an ion beam formation is not related to a material sputtering, and a sputtering of both metallic and dielectric material can be realized when a device compensating an accumulation of positive potential on the target surface is employed. When the target material is sputtered by ions, it may become ionized and additionally accelerated if an additional bias potential is applied to a substrate. A discharge plasma, which is concentrated inside an ion source, helps to avoid a strong heating of substrate material.

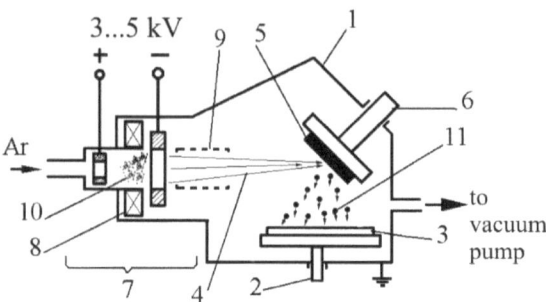

Figure 14: A scheme of an ion-beam sputtering: *1*-a vacuum chamber; *2*-a substrate a holder; *3*-a substrate; *4*-an ion flow; *5*-a sputtered material; *6*-a target holder; *7*-an ion-beam source; *8*-a magnetic system designed to concentrate a glow discharge plasma; *9*-a device intended to focus an ion beam; *10*-a zone of glow discharge plasma concentration; *11*-a particle flow deposited to the substrate.

A disadvantage of ion-beam sputtering is a difficulty to adhere a precise chemical composition of a deposited coating. When an ion impacts a target surface, a great number of complex processes runs in it (a sputtering, a mixing, a radiation-stimulated diffusion a segregation, and a Gibbs adsorption). They can change a chemical composition of the top target layer and the sputtered material [56, 57]. The ion-beam sputtering found its application, in particular, in a fabrication of multilayer laminated structure for a nanoelectronics when a layer thicknesses ranged within 1 to 10nm [57].

An ion implantation. This method is based on a high-energy ion implantation into a material surface. The process is realized in 10^{-4} to 10^{-3} Pa vacuum with the help of an ion-beam accelerator (an implanter). The device (Fig. **15**) is assembled of one or several ion sources, which transform a material to an ionized plasma.

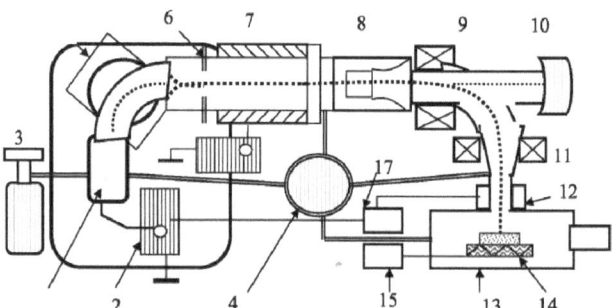

Figure 15: A device applied for an ion implantation: *1*-an ion source; *2*-an ion source feeding block; *3*-a system of gas feeding to a source; *4*-a differential vacuum pumping-out; *5*-an ion beam magnetic analyzer-separator; *6*-an ion beam forming aperture; *7*-an acceleration section and a block of high-voltage sources; *8*-electromagnetic lenses; *9*-a device for a filtration and a beam deviation; *10*-a chamber for an absorption of atoms, particles and impurity ions; *11*-a scanning device (plates deviating a beam along *X* and *Y* coordinate); *12*-a Faraday cup for measurements of beam parameters; *13*-a target reception chamber; *14*-a target holder with a heater; *15*-a device for a heating and controlling of a target temperature; *16*-a vacuum lock with a cassette mechanism for a reception chamber reloading; *17*-a device of a control for ion beam parameters.

Implanted ions are formed by an electric-arc, a thermal evaporation (including a laser one), which are aligned with a glow discharge. Generated ions arrive to a system of analysis and separation, in which those of a non-desired impurity are separated from a primary beam. The ion beam is concentrated using focusing lenses after a separation, and arrives to an accelerator, in which ions are accelerated to high energies in an electric field. A system of electric lenses and deviating plates is envisaged for a stabilization and scanning of the ion beam. Basic parameters of the ion implantation technological process are an energy of accelerated ions E_0 and a irradiation dose D.

Ions are implanted into a modified material and reach 5 to 500 nm depth, depending on their energy. Conventionally, a low-energy ion implantation (2keV to 10keV ion energy) and a high-energy implantation (10keV to 400keV ion energy) are distinguished. An ion beam spot diameter at a treated material surface may be from 10 mm to 200 mm and an average ion current may range from 1 to 20mA depending on an implanter construction [55-58]. A value of the ion irradiation dose is usually $10^{14}cm^{-2}$ to $10^{18}cm^{-2}$. In addition to ion implantation itself (an ion penetration into a material surface), the following processes take place in a material surface: a sputtering, a development of impact cascade, a cascade (ballistic) atomic mixing, a radiation-stimulated diffusion, a metastable phase formation, a radiation-stimulated segregation (a redistribution of material atoms over the surface layer), a preferential sputtering, a Gibbs adsorption (a change of the surface content due to a decrease of free energy), a heating, *etc.* A ratio of these processes depends on a kind of implanted ion, a material, and treatment technological conditions.

Main advantages of the ion implantation are [58]: a possibility to form practically any combination of materials, an increased limit of a solid phase solubility of the system components (*i.e.* such alloys, which are impossible under usual conditions due to a thermodynamical limitation can be formed), a low temperature of a modified material, an absence of notable changes in a sizes, a structure, and properties of a basic material, an absence of a distinct interface, of an adhesion problem, an easily controlled treatment depth, a good reproducibility and stability of a process, a high purity, a possibility to form a complicated surface nanostructure due to a high scanning accuracy of the ion beam over the treated surface, and a possibility to realize the ion implantation simultaneously or subsequently.

Disadvantages of this method are: a material surface can be treated only a region of direct ion beam action, a low ion penetration depth into a material (especially when an energy is low), a sputtering process occurring in a surface, and a high cost.

Laser methods. These methods are employed to form a nanostructured state in a thin surface layer of an ordinary metallic material or a tool using a high-density laser emission [56, 57, 59]. An energy density reaches 10^3 to 10^{10} Wt/cm^2, and a pulse duration is up to 10^{-2} to 10^{-9} s. In a number of cases, a continuous CO_2 laser emission of 10^5 to 10^7 W/cm^2 energy density was used. A beam scanning rate provides 10^{-3} to 10^{-8} s interaction time. A surface material layer of 0.1 to 100 μm was melted very quickly and then solidified with 10^4 to 10^8 K/s rate under an action of laser emission. Due to a short-time thermal action, a total mass of metallic material has no time to be heated providing a high rate of heat removal. A high cooling rate allows a formation of a nanocrystalline and even an amorphous structure. In the latter case, the nanocrystalline state can be formed by an additional controlled crystallization occurring under the thermal treatment. The laser doping or the laser implantation means an additional matter introduction into a melted surface layer. Such implantation may be realized both as a preliminary thin film deposition to a treated material surface, and as an injection of a powder particle (including a nanoparticle) by a gas jet over a laser emission zone. A doping may follow two aims: (a) to form a modified layer, which chemical composition and properties differ from a basic metal, and (b) to make a formation of a nanostructural or an amorphous state easier when a melted surface layer is solidified.

A molecular-beam epitaxy. A molecular-beam epitaxy (*MBE*), in fact, is a thin film vacuum deposition, which was developed to its perfection. It differs from a classical technology of vacuum deposition by a higher control level of a technological process. The *MBE* allows a formation of a thin single-crystalline layer at a heated single-crystalline substrate due to a reaction between a molecular or an atomic beam and a substrate surface, Fig. **16**. A high substrate temperature leads to an atomic migration over its surface. As a result, an atom occupies a strictly definite position conditioning an oriented growth of a crystalline film. A success of an epitaxy process depends on a relation between a film and a substrate lattice parameter, correctly selected conditions of an incident beam intensity and a substrate temperature. When a single crystalline film is growing on a substrate of a certain material without a chemical interaction, this process is called a *heteroepitaxy*. When a substrate and a film material does not differ or differ a little, a process is called *a homoepitaxy or a autoepitaxy*.

The oriented growth of film layers, which start a chemical interaction with a substrate matter, is called *a chemoepitaxy*. A film-substrate interface has the same crystalline structure, but a different composition. In comparison with other technologies, which are used to grow a thin film and a multilayered structure, the *MBE* is characterized, first, by a low growing rate and temperature.

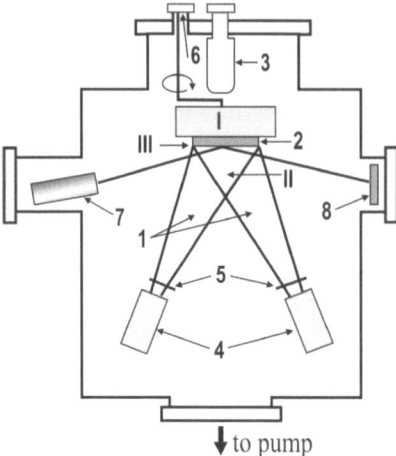

Figure 16: A scheme of a device for a molecular epitaxy: *1*-a sample holder with a heater; *2*-a sample; *3*-a mass-spectrometer; *4*-diffusion cells; *5*-gates; *6*-a manipulator; *7*-an electron gun (DRFE); *8*-a luminescence screen.

An advantage of this method is that a molecular flow of any material can be sharply broken off, subsequently recommenced, and delivered to the substrate surface. It is most important for a multilayered structure with a pronounced layer interface. A possibility to analyze a structure, a composition, and a morphology in the process of layer growth promotes also a formation of perfect epitaxial structure. For these purposes, a diffraction of reflected fast electron (*DRFE*) and an electron Auger-spectroscopy (*AES*) are employed.

An evaporated material is deposited to a substrate under conditions of a superhigh vacuum. The sample is fixed in a manipulator equipped with a heating device, using effusion cells (an effusion is a slow gas leaking through a small hole). The effusion cell is a cylindrical reservoir, which is fabricated of a pyrolytic boron nitride or a high-purity graphite. A heating spiral, which is made of a titanium wire, and a heat screen, which is fabricated of a usual titanium foil, are arranged above a melting pot. An effusion cell can function up to $1400°C$ temperature and sustains a brief heating up to $1600°C$. To evaporate a refractory material, it is heated by an electron bombardment. A temperature of an evaporated material is controlled by a tungsten-rhenium thermal couple, which is clasped to the melting pot. An evaporator is fixed to an individual flange with an electrical lead feeding the heater and the thermal couple. As a rule, there are several evaporators in one chamber. Basic film components and doping materials are set inside every evaporator. The growing chamber of a modern technological complex is usually equipped by a quadrupole mass-spectrometer, which can analyze a residual atmosphere in the chamber and to control an element composition in the course of a technological process. A diffractometer for reflected fast electrons is located in the chamber to control the structure and the morphology of a growing epitaxial structure. It is equipped with an electron gun, which forms a highly focused electron beam of 10keV to 40keV. The electron beam falls to a substrate plane at a very small angle, and a scattered electron wave shows a diffraction image in a luminescent screen.

6.5.2. Methods of Chemical Deposition from Vapor Phase (CVD)

A metallic compound in a gaseous state is deposited to a heated tool surface [60, 61]. As a rule, the deposition is performed in a special chamber at a low pressure, with the help of reduction pyrolysis. In a number of cases, an interaction of a basic gaseous reagent with an additional one is employed. A carbonyle, a halogen, and a metal-organic compound is used most often. For example, a metallic halogenide is reduced to a metal using a hydrogen and form a halogen-hydrogen compound. A carbonyl is decomposed into metal and carbon oxide using the pyrolysis. An optimum temperature for these chemical reactions is $500°C$ to $1500°C$. Therefore, a tool should be heated to such temperatures in order to localize the chemical reaction near its surface and provide an optimum course of the process, high coating properties, and a good adhesion. The coating is formed by a sequential layer-by-layer deposition. The high temperature may also activate a solid-phase and a gaseous-phase element diffusion between the coating and the substrate. A coating of 1μm to 20μm can be formed with 0.01μm/min to 0.1μm/min rate. This method can be employed for a deposition to an inside tube and hole surface. A boron, boride, carbon, carbide, nitride, oxide, silicon, and silicide film can be formed in addition to metallic films [62]. A main disadvantage of the *CVD* is a need to heat tools to a high temperature. On one hand, this negatively influences mechanical properties and a substrate structure. On the other hand, additional problems arise with a formation of nanostructured coatings.

A glow-discharge plasma deposition. The processes is organized in a chamber under low pressure according to the above-mentioned scheme of the cathode or magnetron sputtering, or an ion cladding. There are two varieties of this method. In the case of *reactive sputtering,* ions of a target material interact with ions of an active gaseous medium forming a glow-discharge plasma, which, is deposited to a treated tool surface. A typical example is a titanium nitride coating, which is fabricated by an interaction of nitrogen and titanium ions.

The second variety is called *an ion-activated chemical deposition from a vapor phase.* In this case, chemical reactions are similar to those employed by the *CVD method*, but an activation by the glow-discharge plasma decreases a temperature to 200°C-300°C. Such approach allows one to overcome the above-mentioned disadvantage of *CVD* method. However, a coating does not have a chemical composition of very high purity since a substrate desorption is insufficient at a low temperature. Impurities of a reaction gas may penetrate into a coating. A film of 1nm thickness and thicker can be formed by one of the varieties of CVD method, which is called an atomic layer-by-layer deposition (ALD) [63].

An atomic layer-by-layer deposition employs a principle of molecular assemblage of a material gaseous phase. The deposition process of 1 Å film is realized in several steps: several pulsed gaseous-phase reactions run for a short time period). A deposition temperature is 200°C to 400 °C. Films from 1 nm (and even 1 Å) to several μm thickness can be formed by the *ALD* method. But a typical film thickness is 10nm to 100 nm. This method allows a formation of a nitride, oxide, metallic, semiconductor film, and a nanolaminate with an amorphous or/and a crystalline structure, depending on the deposition temperature. An advantage of this method is a possibility to control a film thickness (with an accuracy of one atomic layer), to form a uniform coating over a large area (a deviation from a uniformity of ≤ 0.25 %), and to realize a uniform deposition to a porous or a bulky material, or a surface of a complicated geometrical form. A resulting film is characterized by a low defect content, a low internal mechanical stress, and is free of a microporosity. The deposition uniformity over the surface of the complicated geometrical form can hardly be reached by other methods like the *CVD* and the *PVD*.

6.6. METHODS OF FULLERENE AND NANOTUBE FABRICATION

A carbon nanocluster-a fullerene can be formed in an arc-discharge using a laser ablation or a catalysis using a transition metal cluster. A classical method of the fullerene formation is a carbon evaporation under vacuum conditions. A carbon vapor is overheated to 10^4 K [63]. Then, an overheated vapor is intensively cooled by an inert gas jet (for example, helium), and a powder with an essential amount of clusters (molecules) is formed. The powder contains two types of clusters of molecules: with a small grain size and an odd number of carbon atoms (to C_{25}), and a big size and an even number of carbon atoms (C_{60} and C_{70}). Then, one group is separated from the other employing, for example, methods of a powder metallurgy since a cluster of the first group is not stable. Also, a molecule with a high amount of atoms (C_{100} and more) can be obtained selecting the process parameters. There is a number of other methods [64-70].

A thermal evaporation. To realize this method, an ohmic heating of a graphite rod in helium atmosphere at $P = 100$ torr is employed. A carbon condensate is accumulated at a glass disk. Then, this black powder is scraped from the disk and embedded in a benzene. A resulting suspension after drying looks like a dark brown or an almost black material. CS_2CC_{l4} can be employed instead of the benzene. A relative output of C_{60} can be essentially increase with this suspension. About 1g of C_{60} can be formed for 24 hours. The benzene dissolved the fullerene in a total bulk. When the benzene was dried, the fullerene turns out at a soot particle surface, which was employed to increase the fullerene yield under irradiation.

An arc method. A scheme, which is used to form a fullerene, is shown in Fig. **17**. One electrode is a flat disk, another one is a sharp rod of 6mm diameter, which is pressed against the first electrode by a spring.

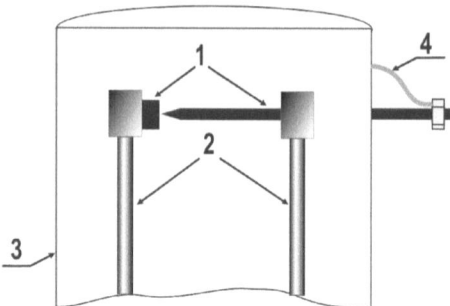

Figure 17: A scheme for a fullerene fabrication by an arc method: *1*-graphite electrodes; *2*-a copper cylinder with water cooling; *3*-a water-cooled surface, at which a coal condensate is precipitated; *4*-a spring.

A collecting surface is a copper cylinder with water cooling of 8cm diameter and 15cm length. Helium is a buffer gas at 100torr pressure. An alternating current of f = 60Hz, I = 100A to 200A, U = 10V to 20V was conducted through the electrodes. A graphite was evaporated with a rate of 10 g per hour and an efficiency of fullerene fabrication was 1g per hour C_{60}/C_{70} = 10/1 when the electrode pressure was optimal and weak. In a certain time, the soot was scraped and placed in a boiling toluene for 3 hours. A resulting dark brown liquid was evaporated using a rotating evaporator. C_{60} dominated when both electrodes were sharp (I = 100A to 180A, U = 5V to 8V, P_{he} = 180torr), but a fullerene fraction was lower ~ 50 mg per hour.

A carbon nanotube synthesis. An arc-discharge and a laser evaporation of a carbon target are successive methods for a nanotube synthesis. A heat-induced decomposition of a carbon material by a catalyst, which is used to form a carbon nanotube (*CNT*), seems to be the most suitable method to realize a large-scale synthesis. The carbon nanotube may be grown under astonishingly different conditions. There are two types of nanotubes: a *MWCNT* (a multiwall carbon nanotube), which is grown without a catalyst and *a SWCNT* (a single-wall carbon nanotube), which is grown only with a catalyst. In contrast to a long catalytically grown fiber, when the fiber ends are usually decorated by catalyst particles, the *SWCNT* ends are closed without traces of the catalyst. The nanotube looks like a rod, which grows with a rate of about 1μm per minute at a cathode surface. Optimal experimental conditions for a multiwall nanotube growth are: about 20V voltage between the electrodes, 150A/cm^2 current density, 500torr helium pressure in a chamber, and a constant inter electrode gap of about 1mm. In general, an anode diameter should be lower than that of a cathode, and both electrodes should be efficiently cooled by a water. A temperature of an interelectrode gap should be close to 3500 °C. A modern automatic generator, which is used to form a carbon arc, is equipped by an optic-electron control for a graphite electrode configuration and a spectroscopy for a plasma diagnostics. In this case, the best yield of carbon nanotube and nanoparticle is approximately about 25 wt.%. The first single-wall nanotube was formed by an arc-discharge using the catalyst and an evaporated carbon. For this purpose, a hole was drilled in the anode center point, through which a mixture of a metallic catalyst and a graphite powder (a metal fraction was 1wt.% to 10wt.%) was filled inside. Several catalysts were tried, but the best "harvest" of nanotubes was picked-up with Ni, Co, and a bimetallic system like Ni-Y, Co-Ni, Co-Pt. Formations contained a great amount of junctions with 10 to 100 *SWCNT*s, an amorphous carbon, and catalyst atom nanoparticles. The application of Ni-Y catalyst (4:1 proportion) and arc-discharge resulted in a very high quantity (> 75 %) of *SWCNT*s. A semicontinuous procedure of SWCNT synthesis using a hydrogen arc discharge and a mixture 2.6at.% Ni, 0.7at.% Fe, 0.7at.% Co and 0.75at.% FeS allowed more than 1g of nanotubes per hour.

One more efficient way of the single wall nanotube fabrication is a laser evaporation demonstrated in Fig. **18**. A laser beam (*A*) is introduced into a chamber and is focused at a metallic composite target (*B*) with the help of a mirrors system. An inert gas is fed through a hole (*C*). A product is collected in a system of copper wires positioned inside a quartz tube (*D*), which is snapped-together with a filter and a device for a gas pumping-out. A direct laser evaporation of a composite (a transition metal-a graphite), for example Co-Ni, 1% Ni-Y (4.2 : 1 %; 2 : 0.5 %), which is an electrode target placed in a helium or argon atmosphere of a furnace, has > 80 % efficiency.

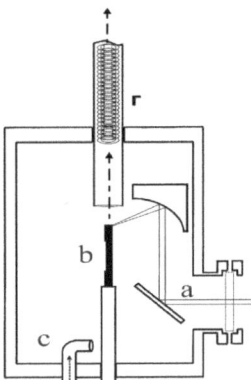

Figure 18: A scheme for a fabrication of a singlewall carbon nanotube by a laser technology.

The furnace is heated approximately to 1200°C (a continuous emission of 250W CO_2 laser of 10.6μm wavelength does not need the furnace). An amount of carbon, which is deposited as a sooth, can be decreased by an action of two subsequent laser pulses: the first one is for an ablation (a reduction and a heating) of the carbon-metal mixture, the second one is purposed to destroy big particles in the ablation zone and to use them to grow a nanotube structure.

SEM images for *SWCNT* bundles after the laser ablation with the catalyst Ni:Y (2 : 0.5at.%) are presented in Fig. **19**.

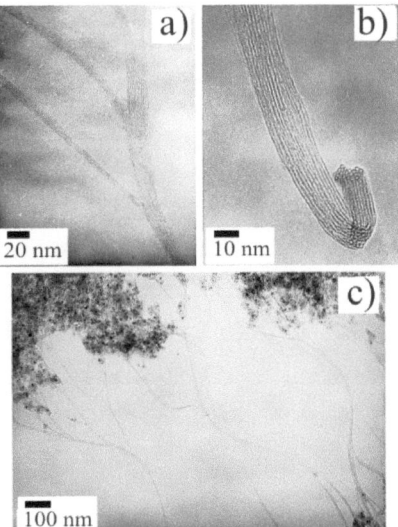

Figure 19: A SEM image of a SWCNT bundle, which is formed with Ni:Y catalyst (2 : 0.5at.%): *a*-a formation of bundles from individual SWCNTs; *b*-a bundle cross-section; *c*-a formation bigger structures from bundles.

Fig. **19** shows how an individual *SWCNT* forms a bundle. The laser ablation and arc-discharge with NiCo catalyst demonstrated high *SWCNT* output. It has the following succession: Ni-Co > Ni ~ NiFe » Co ~ Fe > Pd ~ Pt. A low catalytic ability of Pd and Pt as graphitization catalysts causes a low catalytic activity of *SWCNT* formation. Due to a high Fe solubility in carbon, a segregation temperature of Fe particles from C-Fe solution is low. Therefore, the *SWCNT* output with Fe-catalyst is low. An unstable crystalline Co orientation in a graphite is probably an obstacle preventing the SWCNT growth, since the crystalline Co orientation changes very often under conditions of catalytically grown SWCNT. As an alternative, this could give rise to a fast increase of Co particles in C-Co mixture at a high temperature, when the SWCNT is unstable. NiCo, Ni, and NiF are very efficient graphitization catalysts. They have a low carbon solubility and a stable crystalline phase and orientation in a graphite. A great amount of SWCNT can be formed using a mixture of Ni and Y (4 to 6 : 1) and the arc-discharge method.

Recently, many publications reported about a fabrication of a single wall and multiwall CNT with a catalyst seed (Fig. **20a**) or a floating catalysts (Fig. **20b**) using a chemical vapor decomposition (CVD) of a carbon containing material. A nanotube diameter can be varied by changing a size of an active particle in the catalyst surface.

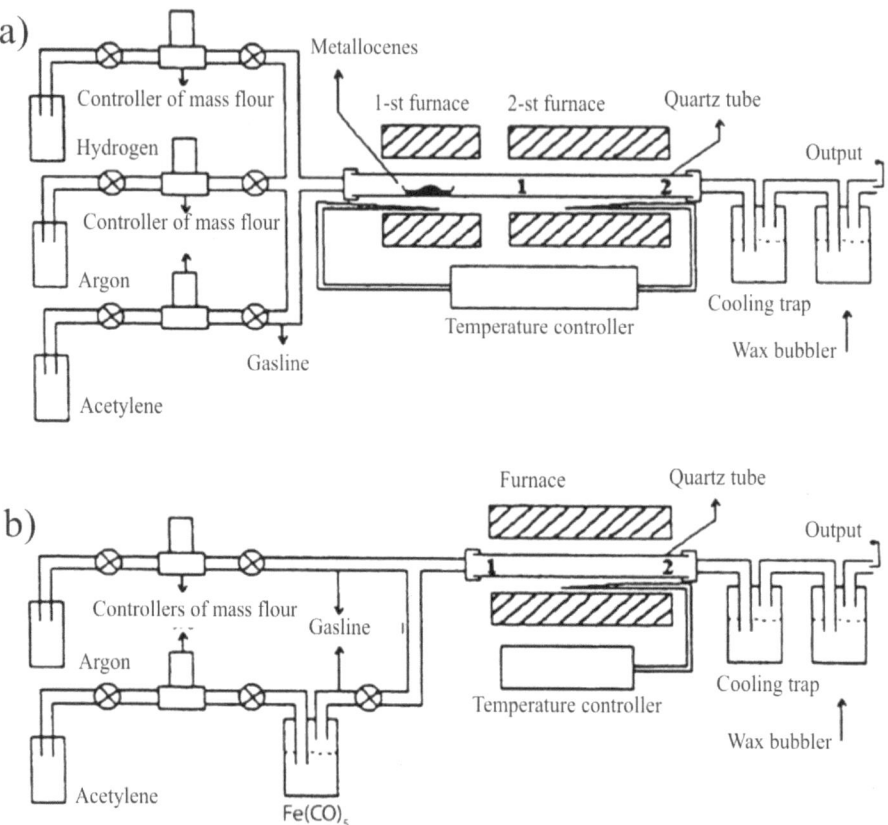

Figure 20: A pyrolysis system employed for a SWCNT synthesis using a CVD method: a-a metallocene; b-Fe(CO)$_5$ with an acetylene.

The nanotube can be grown with the help of various catalysts embedded into a substrate, and at various temperatures from 500°C to 1200°C. A resulting layer may contain a high amount of the single wall and the multiwall CNT with 1.5 to 20 nm diameter. Usually, they are organized as a bundle of less than 100 nm size and several mm length. A summary quantity of SWCNT in such 40 nm bundle can reach 600 or more.

The method based on the floating catalyst has a higher productivity in comparison with the embedded catalyst and can ensure a continuous CNT growth. A hexane, an acetylene, a high pressure CO (HiPCO), and a xylene are most popular as carbon containing materials. A ferrocene Fe(CO)$_5$ is a catalyst-containing organic material. It exists as a gaseous phase and is an additional carbon source.

A direct synthesis of a long bundle of an ordered SWCNT using an optimized technique (CCVD) is realized in a vertical furnace with the floating catalyst. A hexane solution with a given ferrocene (0.018g · ml^{-1}) and tyrophene (sulphur addition 0.4 wt.%) proportion is fed with 0.5 ml · min^{-1} velocity into a reactor. Preliminarily, the reactor is heated to the pyrolysis temperature (1150°C) and a hydrogen is pumped-inside with 250 ml · min^{-1} velocity.

This growing procedure is continuous and allowed a *SWCNT* output of 0.5 g per hour. The formation of very long *SWCNT* bundles (to 20 cm) is a unique feature of this vertical way of *CNT* growing (Fig. 21). As a rule, the *SWCNT* formed by the ferrocene-assisted method of a hydrocarbon decomposition (a benzene, a xylene) grows at $T \sim 1050K$, and a mix of the single-and multiwall nanotubes is formed at high temperatures (> 1300K).

A thiophene promoted an increased *SWCNT* output. A tube bundle contains impurities (~ 5 wt.%) with Fe-catalyst particles and amorphous carbon like it is with the other ways of *SWCNT* fabrication. In general, the bundle is about 0.3mm diameter, and it is thicker than a human hair. A high-resolution *SEM* image of one "rope" of bundles indicates that bundles of directed *SWCNT* are organized as a two-dimensional triangular lattice (Fig. **21b**). The *SWCNT* diameter varies within 1.1nm to 1.7nm.

A "thrilling" growth of various tree-like carbon micron-size structures is observed, when a temperature of a hot graphite surface ranges between 1100°C and 2200°C. In this case, the CVD method was employed for their fabrication from methane without any catalysts. A carbon deposition performed under specific conditions (for example, an application of quick heating and cooling cycle) may stimulate the formation of structures with a very unusual morphology.

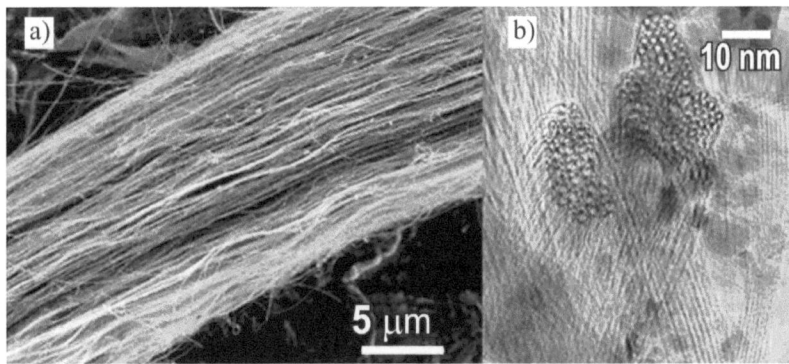

Figure 21: A micrographic image of a typical "rope" of single-wall carbon nanotubes: *a*-a *SEM* image for a "rope" weaved of thousands of nanotube bundles (*D* = 1.1 to 1.2 nm); the "rope" diameter is about 15μm; *b*-a high-resolution *SEM* image for the *SWCNT* bundle cross-section, which illustrates a two-dimensional triangular lattice.

All procedures of a *CNT* purification follow definite stages: a preliminary filtration is purposed to get rid of big graphite particles; a dissolution is to remove fullerenes (occurring in an organic solvent); and the catalyst particle (occurring in a concentrated acid), a micro filtration, and a chromatography are employed to distinguish an individual MWCNT and nanoparticles or a *SWCNT* and an amorphous carbon cluster (Table **2**). As an example, in brief, let us discuss the *SWCNT* purification procedure in the case when it is formed by CO decomposition under high pressure (Table **2**). An initial low-density HiPCO tube is pressed-in into a dry filter by depositing *SWCNT* to the filter holder under vacuum conditions.

Table 2: A weight loss and a metal concentration after every purification step.

Sample	Metal %	Weight loss
(*a*) initial SWCNT	5.06	
(*b*) initial SWCNT heated under 225°C in damp Ar/O$_2$ for 18 hour	0.67	33.7%
(*c*) heated under 325°C in damp Ar/O$_2$ for 1.5 hour	0.05	8.3%
(*d*) heated under 425°C in damp Ar/O$_2$ for 1 hour	0.03	22.9%
(*e*) annealed in Ar under 800°C for 1 hour	0.03	4.2%

- Every oxidation step is accompanied by a treatment in a concentrated HCl-acid solution for 1-15 min. Then, a nanotube is filtered and dried in a vacuum furnace at 100 °C for not less than 2 hours.

- A metal % = Fe %, which is calculated from measurements of a gravimetric analysis as Fe/(C + Fe) atomic per cent.

- A total weight loss = 69.1 %. A total weight loss without annealing at 425 °C is 46.25%.

Vacuum conditions help to concentrate these light objects near a filter holder. The *SWCNT* (usually ~ 100mg) is placed into a ceramic "boat" and inserted into a tube of a quartz furnace. A gaseous mixture of 20 % O_2 and Ar (or air) in the form of gas bulbs is leaked through a water and a sample with a full 100sccm velocity. The nanotube is heated to 225°C for 18 hours, which is accompanied by an ultrasonic treatment during 15min or a long-time treatment (for a night) in a concentrated HCl solution. Then, the tube in the acid solution are filtered through 1.0 µm pores of a Teflon membrane of 47 mm diameter and washed several times in a deionized water and a methanol. Further, they are dried out in a vacuum bake oven at 100 °C for minimum 2 hours and weighed (using a thermogravimetry). An oxidation and an acidic extraction cycle are repeated at 325°C for 1.5 hour and at 425°C for 1 hour. The purified tube is annealed at 800 °C in Ar-medium for an hour after drying in a vacuum bake oven. A typical weight loss and a metal concentration after every purification stage are presented in Table **2**. The weight loss, as it is seen, increases dramatically at 325°C to 425°C. A metallic particle is closed from a carbon material (a shell) if a temperature step of oxidation in a damp Ar/O_2 atmosphere (or a damp air) is low. Evidently, a carbon shell is destroyed and metallic particles are transformed into an oxide and/or a hydroxide. An expansion (densities for Fe and Fe_2O_3 are 7.86 and 5.18 g/cm^3, respectively) of a metallic particle due to a low oxide density results in a fracture of a carbon shell, its opening, and a metal release. This phenomenon depends on an HCl capability to release an iron from an initial material only under conditions of the damp oxidation in Ar/O_2 medium. The enucleated metallic particle further catalyzes other forms of carbon and the *SWCNT* after removing of the carbon shall.

Formation of arrays of oriented carbon nanotubes and their fine architectures. An oriented nanotube may be formed from a carbon-containing material using various methods: a thermal activation or a plasma reagent excitation (with or without a catalysts), a nanolytography, etc. A synthesis of the oriented *CNT* is a very promising way in comparison with a laser or an arc method, since a synthesis product is free from a non-desired carbon nanostructure. As it was mentioned above, a chemical, a physical, and an electron *CNT* property depends on a geometry and a structure, which are specified by a preparation procedure. Not long ago, a regulation of such parameters as: a catalyst dimension and type, a reaction gas pressure, and a temperature allowed a formation of *CNT* arrays owing various orientations and structures. An interlaced *MWCNT* is synthesized at SiO_2 substrate around a cobalt nanoparticle. A well-oriented *MWCNT* is formed around a porous quartz doped by iron particles, around a cobalt-coated silicon, a nickel and an iron coated glass, Al_2O_3, Si_3N_4 or a silicon using the thermal *CVD*-method or the plasma-stimulated *CVD*. The *SWCNT* is grown by a decomposition of a hydrocarbon gas, such as CH_4 or CO, around a many-atomic metallic catalyst Fe, Co, Ni, and Fe-Mo and Co-Mo alloy.

A catalyst type strongly affects a nanotube diameter, growing rate, wall thickness, morphology, and microstructure. A nickel ensures the highest growing rate, the biggest diameter, and the thickest wall, while Co results in the lowest growing rate, the smallest diameter, and the thinnest wall. The *CNT*, which is formed with Ni catalyst, has a strong orientation, the smoothest and purest wall surface. An amorphous carbon with nanoparticles in its external surface covers a tube grown with Co catalyst. As a rule, a finer film of metal-catalyst results in a formation of smaller catalyst particles and, therefore, a nanotube diameter is smaller. A growing process and a structure of the *CNT* strongly depends on a temperature. The temperature effect on the *CNT* growing process and structure under conditions of *CVD* was studied for the case, when an iron was incorporated into a quartz at various temperatures ranging from 600°C to 1050 °C and gas pressures amounting 0.6torr to 760torr. The *CNT* grown by *CVD* at a low gas pressure and temperature is completely hollow and looks like a bamboo. The *CNT* diameter essentially increases with the temperature. The tube diameter is larger at low gas pressure since an amount of graphite layers increases in the CNT wall. Whereas the diameter is larger at high pressure, since an amount of wall graphite layers and the CNT inside diameter increase simultaneously. These results indicate that a growing temperature is a critical parameter of the CNT synthesis [52-59].

Fig. 22 shows a typical experimental system, which is applied to form an oriented CNT with the help of a floating catalyst.

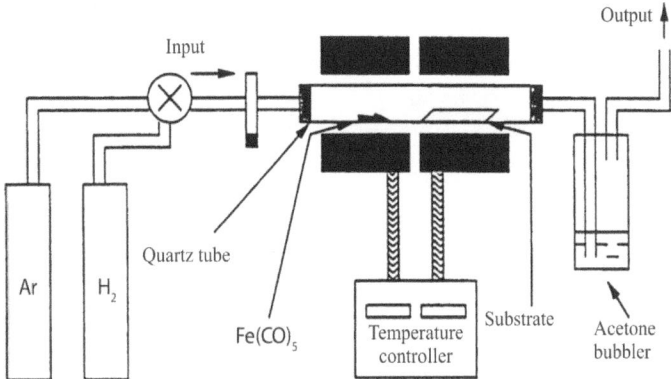

Figure 22: An equipment employing a pyrolysis for a synthesis of an oriented CNT with a floating catalyst.

A large area vertically oriented CNT can be synthesized by the pyrolysis of an iron phthalocyanine $FeC_{32}N_8H_{16}$ (Fe, Pc) in Ar/H_2 flow at different substrates at 800°C to 1100 °C in a double furnace. An array of oriented CNT may be grown at a partially masked/preliminarily structured surface or using a contact printing. A carbon nanotube grown at a flat microstructured Si substrate is illustrated in Fig. 23. The process is realized by the CVD at 800 °C using an acetylene pyrolysis. Here, Fe-films of 30nm thickness, 5μm x 5μm, 10μm x 10 μm, and 20 μm x 20 μm sides, and 15 μm, 20 μm, and 30 μm distance between rows, respectively, is deposited to Si surface. It plays a role of the catalyst in the process of CNT growing. Then, a microstructured system of vertically oriented carbon nanotubes is formed at Si flat surface using the CVD method. A vertical arrangement of carbon nanotube is provided by a Van der Vaals interaction between adjacent CNTs. The CNT deviation from a vertical position demonstrated in the figure is observed at the points, where the Van der Vaals force fails to provide the oriented arrangement (the edges of the rows). A process of CNT fabrication, which is described here, is very promising for a production of radio-frequency amplifiers, electron guns with a cold electron emission, and displays, which are employed for various purposes.

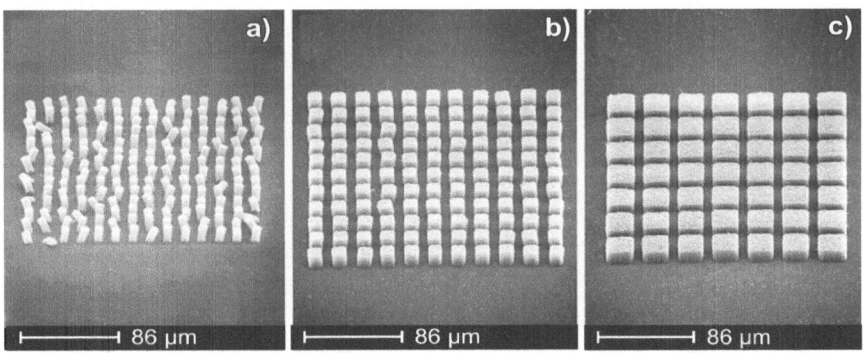

Figure 23: A CNT grown at a flat microstructured Si substrate: a-a square structured Fe film with 5μm x 5μm side, and 15 μm distance between the sides (left); b-a square region of 10μm x 10μm side and 20 μm distance between cells (center); c-a band of 20μm x 20μm side and 30μm distance between cells (right).

A gate electrode is used to govern an emitted electron flow. In this case, a simple and an advisable device is a vertical diode purposed for a field emission and containing a carbon nanotube emitter in a groove region (Fig. 24).

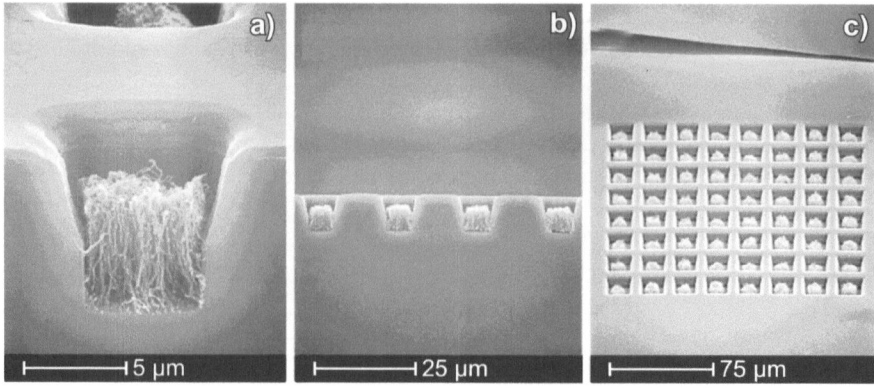

Figure 24: A *SEM* image of a groove of 10μm depth with a carbon nanotube, which is grown only in a bottom: *a*-a side view; *b*-a top view.

The groove is purposed to form a triode structure based on the *CNT*. This process includes a deposition of several thin film layers, their structuring (a photolithography), an etching, and a deposition of Fe-catalyst to the groove bottom. The groove is prepared in the following way: first, several thin films are built-up, microstructured, etched, then, Fe catalyst is deposited to the groove bottom [57-59].

In spite of a high control for the *CNT* growing process, which exists in the *CVD* and the plasma-enhanced *PE CVD* methods, as a rule, they need a treatment at temperatures above 500°C. This essentially limits materials, which could be selected as the substrate and for an integration process. The *CNT* fabricated for a production of a flat-panel display and in a vacuum microelectronics, should be deposited to a glass or a polymer base at not lower than 300°C temperature. A vertically oriented *CNT* grown on a plastic substrate (a kapton polymer film and silicon), which are employed for a field emitter, are be fabricated at a low-temperature (120°C) using *PECVD*. A measured emission yield demonstrated a low field of emission activation, *i.e.* a field, for which $j \approx 10^{-9} A/cm^2$ (3.2 V/μm), and a low threshold field (4.2 V/μm), *i.e.* field under which $j \approx 10^{-6} A/cm^2$.

Another promising way to govern the oriented *CNT* growth is a synthesis at a template (a profile). When the *CVD* is formed in a template pore (of several nm diameter), it is limited within this pore. Then, the template is dissolved or etched, and many free staying oriented *CNT* remain at hand. A polymer, a metal, a semiconductor, or another material can be deposited to the template pore. In principle, almost any hard material can be synthesized in the template nanopore. There are several methods of a nanostructure synthesis in a template: an electro-chemical deposition, an electrolytic deposition, a chemical polymerization, a sol-gel deposition, and the *CVD* technology. An employment of the template allows a development of many new optical and electron devices, a *MEMS* (a microelectromechanical system), a biomedical chromatograph, a sensor, a transducer, and an efficient field emitter.

A synthetic mineral crystal, an aluminum phosphate *AlPO4-5*, a quartz aerogel, a mesoporous quartz, a silicon, and an anode aluminum oxide (*AAO*) are materials with a high porosity, which found their application in a synthesis of the oriented *CNT* using the *CVD* method. Today, a synthesis of the regularly arranged *CNT* at a porous silicon and a nanoporous *AAO* is a well developed procedure applied in a number of laboratories. A template based on *AAO* is now most widely used for the nanomaterial synthesis. A self-organized membrane of an anode aluminum oxide fabricated under definite electro-chemical conditions has a porous structure with uniform and parallel-arranged nanopores (Fig. **25**) [59].

A pore diameter can be regulated electrochemically and range from several nanometers to several hundred of a nanometer. This is an ideal material for the template, which is suitable for a fabrication of arrays of oriented nanostructures. A pore density can reach 10^{11} pore · cm^2, and a typical membrane thickness may vary from 10μm to 150 μm.

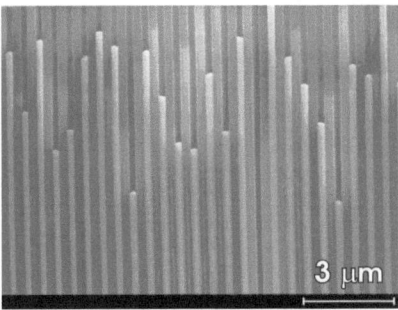

Figure 25: A side view of *SEM* image for channels of 500nm diameter in Al_2O_3 filled by a silver.

The *AAO* (the anode aluminum oxide) membrane may be removed chemically (dissolved) after a termination of the *CVD* process remaining an ensemble of staying free tabular nanostructures, an outside diameter of which is close to a pore size. In such a way, very large panels of well-oriented *CNTs*, which are applied as a flat-panel display with a cold cathode, can be formed using the membrane method.

A soft lithography method employed for a fabrication of a structure-oriented *CNT* array opens a possibility to produce various nanodevices for a wide circle of applications. Growing a three-dimensional architecture of the carbon nanotube, which may be integrated into a micro-electron circuit or a *MEMS* remains still as a future task. One-or two-dimensional junction and/or a transition of nanotubes were fabricated *in situ* in the course of growing and subsequent nanostructuring. However, a nanotube junction to a substrate and/or a metallic film still stays a critical problem for a realization of a three-dimensional device based on *CNT*. An example of a successful realization of such architecture is a *MWCNT* grown under a thin Ni layer, when a nickel film is broken away from Si substrate in the process of CNT formation and remains at the tube edge. In this way, a thin film metallic region is bound to Si substrate by an ensemble of nanotubes and gives a possibility to create the three-dimensional nanotube architecture.

Specialists of the Troy Polytechnic Institute (USA) developed and formed a nanotube structure at a microstructured region around a quartz (SiO_2) and a silicon surface. A microstructuring of Si/SiO_2 system is realized by a standard photolithography with a combined wet and/or dry etching. A template catalyst material is not used in this procedure. A *CNT* of 20 to 30 nm diameter is grown by the *CVD*, which is realized as a substrate exposition to a medium of xylene/ferrocene vapor mixture ($C_8H_{10}/Fe(C_5H_5)_2$) at 800°C. A silicon does not suit for *CNT* growing, however, a vertically oriented nanotube can be magnificently grown at SiO_2-microstructured region (Fig. **26**). Every column is composed of several scores of nanotubes, which grow in a vertical position, perpendicular to SiO_2 cell of Si/SiO_2 template. A simultaneous integration of ordered nanotube structures in various geometrical directions of one common substrate is important for a number of micro-and nanoelectromechanical devices.

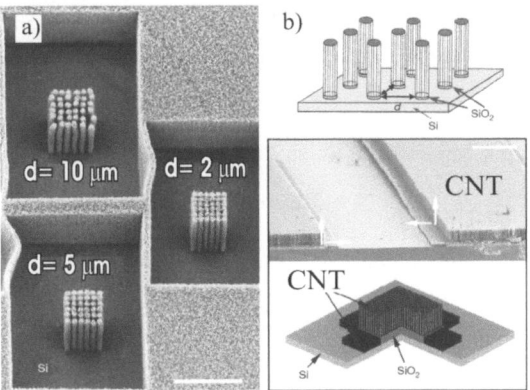

Figure 26: An oriented nanotube: *a*-a *SEM* image of three blocks of cylindrical columns (≈ 10μm diameter) of oriented *CNT* systems; *b*-a vertically and horizontally oriented *CNT* at a cross-section of Si/SiO structured plate (a cells is ~ 100μm).

The fact that the nanotube grows along SiO₂ surface normal means a good possibility to form simultaneously *CNT* arrays with a mutual orthogonal orientation, and even arrays, which are angled one to another, for example, like regions formed at a quartz substrate surface, which are not orthogonal to a basic substrate surface. An electric field also allows a control of the tube directed growing. This effect and a microstructurization of regions with the 6 solid catalyst is compatible to a modern strategy adopted now in a semiconductor technology and may contribute to the nanotube integration into a complicated device architecture.

Not long ago, a controlled synthesis (a catalytic decomposition of acetylene) of a spiral (coil-like) nanotube was developed. An arrays of oriented nanotubes are applied as a template providing conditions for an asymmetric growth of the *CNT*. The spiral nanotube with various coil steps and diameters demonstrated its potential application for nanoelectronics and nanomechanics.

6.7. APPLICATION OF CHARGED PARTICLE BEAM OF LOW AND AVERAGE ENERGY FOR NANOTECHNOLOGIES

A development of future nanotechnologies needs studies of physical, chemical, and mechanical properties of materials and objects with a nanosized structure including a possibility to influence their properties. In this connection, a development of new types of devices, which could help to analyze a structure and an element composition of new nanomaterials and nanoobjects, as well as new technologies, which could be employed for their formation, seem to be a priority. Among a wide circle of physical principles, which could serve a basis for the development of new devices, a special attention is paid to a focused beam of charged particles of a low and an average energy. First of all this is related to the fact that the focused beam ranges from several units to scores of nanometer. Therefore, an interaction of beam particles with a matter can locally modify physical and chemical properties within a nanosize range [62-71].

Today, the focused electron beam, which finds its wide application in a scanning electron microscopy, is also widely applied in a beam lithography (*EBL*, an e-beam lithography) [77, 78]. Due to its ability to sputter atoms of a treated material, the focused beam of heavy ions, which earlier was applied mainly for a secondary ion mass-spectrometry, is now applied for focusing ion beam devices-(*FIB*-the focused ion beam) to form a three-dimensional nanostructure [79, 80]. First, the focused beam of light ions of mega-electron-volt energies was applied for a nuclear scanning microprobe in a local element microanalysis of a matter. Now, an application of the nuclear microprobe popular in an ion beam lithography, which is aimed to form the three-dimensional nanostructure [81, 82]. This technology is named a p-beam writing (*PBW*). A consideration of this technology in this brief report deserves a special attention. Problems arising in a formation of a mega-electron-volt nanobeam and questions relating to a metrology of a beam parameter determination, an application of various resistive materials, and general principles of the *PBW* technology are considered. All aspects are compared with more developed technologies *EBL* (the electron beam lithography) and *FIB* (the focused ion beam technology). For deeper understanding of abilities of nanotechnologies relating to a nanostructure formation, certain application fields are considered: a nanoimprinting, a silicon micro-treatment, a biomedical application, a micro photonics, and microjet devices [68-78].

6.7.1. Features of Accelerated Charged Particles Passing in Matter

Various sorts of charged particles interact in a different way to a solid matter depending on energy. Fig. **8** shows three beam sorts [83]. They are low-energy electron, heavy ion, and light ion beams of an average energy. When the low energy electron beam interacts with a matter, its electrons are mainly scattered at atomic electrons of the matter. As a result, primary electrons many times deviate with a large angle and form a classical pear form in an ionization region around a touch point of the primary focused beam. Fig. **27** shows calculations using a computer modeling by a Monte-Carlo method [84].

The focused electron beam of 50 keV energy penetrated to 40 μm depth into a resistive material (*PMMA*-a polymethylmethacrylate) with to 20 μm deviation from an axis.

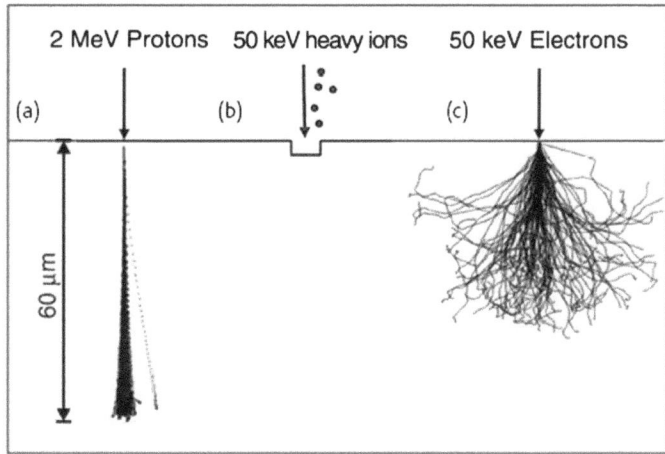

Figure 27: A schematic image of how various kinds of a charged particle beam of low and average energy interact with a matter [83].

This feature of the electron beam hinders a formation of a three-dimensional structure with a high proportion of an object height to its characteristic surface area. Also, a proximity effect, which means a presence of secondary electrons with an essentially high energy component introducing an additional dose in the course of irradiation, is one more negative factor influencing the irradiation process. A correct dose determination is required.

When the focused low energy heavy ion beam interacts with a matter, incident ions are scattered at nuclei of target atoms. The ion beam impulse transferred the material near surface layer reorders atoms inducing a chemical and structural change such as sputtering of atomic and molecular formations. The sputtering process calculated using a Monte-Carlo method (for example, using a numerical code *SRIM* [84]) demonstrates that a sputtering rate of Ga ions of 30keV energy is from 1 to 10 target atoms per an incident ion, depending on a material type. However, the sputtering rate may be significantly increased due to molecules of a chemically active gas (for example a chlorine), which are introduced to a region of a beam-material interaction.

The focused beam of light ions of several megaelectronvolt energies may interact both with electrons and with atomic nuclei of an irradiated material. However, a probability of the ion interaction with electrons is several orders higher than a probability that it would be scattered at atomic nuclei in the first half of its route. The ion-electron interaction cannot crucially change a trajectory of the incident ion, which is hardly different from a straight line, due to an essential mass difference, as illustrated in Fig. **27**. Since energy, which is lost by the ion for the interaction, is low (about 100eV), thousands of electron-atomic interactions is to occur until the ion will fully loose its kinetic energy. When the energy is lost, and, consequently, the motion rate decreases, a probability of atom-nucleus interaction increases and the ion notably changes its trajectory (Fig. **27**). In comparison with an electron beam, the light ion beam of an average energy is characterized by an absence of secondary electrons, an energy of which can essentially contribute to the irradiation dose (the proximity effect). A depth of ion penetration into a material depends on energy and is strictly determined. It is an important property and allows a formation of a multi-level three-dimensional object from a single-layer resistive material. Calculations, which are performed using the numerical code *SRIM*, indicate that 2 MeV-energy proton penetrates into the *PMMA* up to 60.8μm depth with 2 μm deviation at the end of its rout. However, beam expansion in 1μm depth is only 3 nm and 30 nm at 5 μm. This property allows a formation of a three-dimensional nanoobject with a high ratio of the object height to its characteristic surface area and a high wall quality (roughness is about 3nm).

6.7.2. Probe Systems for Charged Particle Beam Formation

A principle of an electron beam formation in a system of an electron beam lithography (*EBL*) are similar to a scanning electron microscopy (*SEM*), where an electron beam is focused as a spot at a sample using electromagnetic lens with a coaxial symmetry and scans a desired region yielding some image. A basic difference between the *EBL* and the *SEM* is that a secondary and a back-scattered electron, as well as a characteristic X-ray emission induced by the beam electron are registered in the case of *SEM* by a detecting device for a purpose to obtain an image and perform an element analysis.

Important points of the *EBL* are the following. A sample can be positioned with a high accuracy due to a control of a laser interferometry. There is a system for measurements of a spot size. Current is measured by a Faraday cup equipped by an integrator, which is able to measure a picoampere current for a dose normalization. There is a high-rate system, which is able to break a beam when a sample is relocated. There is a special program software, which is able to control a beam scanning according to a numeric template. The template is prepared using one of *CAD* programs. Fig. **28** demonstrated in [81-83] shows a scheme of a commercial device JBX-9300FS with a thermal-field electron gun, which is constructed by JEOL Company [83].

A minimum spot area made in a sample surface by a beam is 4nm under 50 pA to 50 nA current. A scanning frequency is to 50 MHz, an approximate scanning areas is 500 x 500μm^2, an accurate scanning area is 4 x 4μm^2 (secondary deviating coils), under 1nm resolution.

A general scheme of formation of a focused heavy ion beam of low energies in an *FIB* device is demonstrated in Fig. **29**.

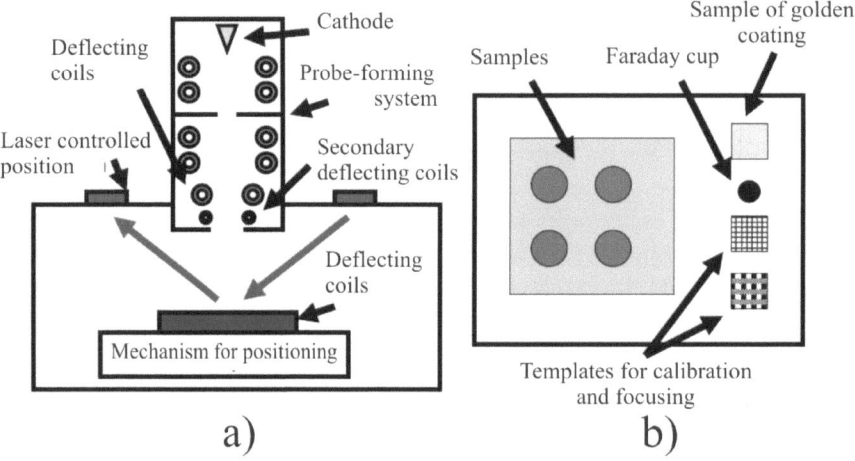

Figure 28: A schematic presentation of JBX-9300FS system: *a*-a general view; *b*-a sample holder and assistant objects.

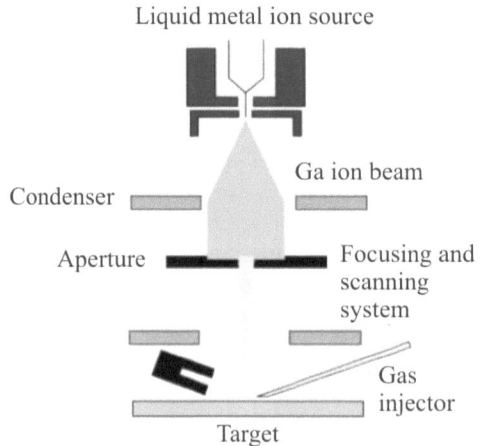

Figure 29: Formation of a focused heavy ion beam of low energy in an *FIB* device.

Here, general principles of the focused beam formation are similar to an electron beam focusing system. A main difference is an application of a gallium ion beam (Ga⁺) instead of an electron one. The ion beam is formed with the help of a liquid-metallic source (*LMIS*), in which an applied high voltage induces an emission of positively charged ions coming from a liquid gallium conical electrode. A set of apertures is used to select a desired beam current and to provide a desired beam focusing area. A typical beam energy is 30 to 50 keV, the best possible resolution is 5 to 7 nm. The beam scans a sample, which is placed in a vacuum chamber under 10^{-7} mbarr pressure. When it is incident to the sample surface, electrons and scattered atomic and molecular formations are emitted. A detection of secondary electrons allows us to obtain an image of the sample surface in a scanning raster. A full description of the operation principle and application is reported in [81].

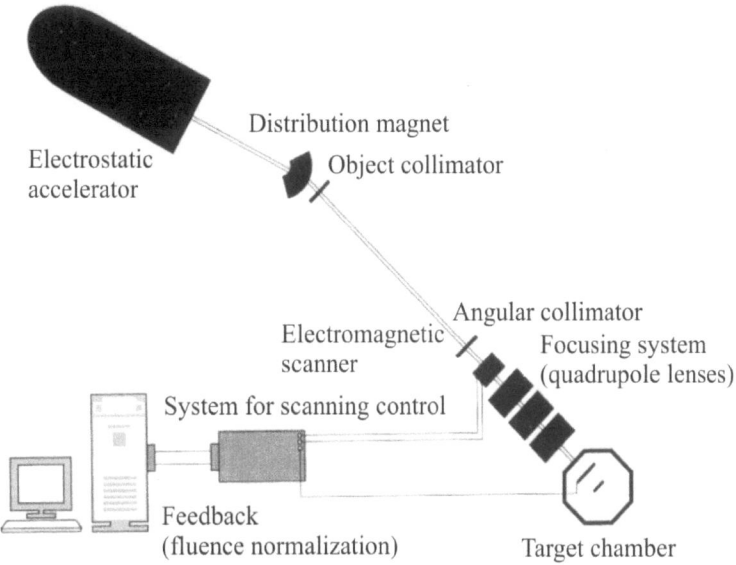

Figure 30: A device employed for a formation of a focused beam of light ions of an average energy using a nuclear scanning micro-probe for p-beam writing.

Similar to an electron beam lithography, a specialized scanning micro-probe for the ion beam lithography is constructed of an ion gun, a probe-forming system, and a target chamber (Fig. **30**). Here, an electrostatic accelerator with an analyzing magnet plays a role of the ion gun. The beam of accelerated light ions (H⁺, H^{2+}, He⁺) of several MeV energy is formed at its exit. Further, it is formed using a system of object-and angular-collimators in order that the focusing system of multiple magnetic quadrupole lens could focus a beam at a target within an area of a minimum possible size. Today, the best results are obtained with a device belonging to the Center of Ion-Beam Applications (CIBA) of Singapore National University [88-90]. An electrostatic accelerator "Singletron^{TM}" of the HVEE Holland Company [91] yields 3.5 MV maximum voltage at a conductor providing a high beam monochromatism of $\Delta E/E = 10^{-5}$ and a brightness about 70 pA ($\mu m^2 mrad^2 meV$). The focusing system is constructed of magnetic quadrupole lens triplet (MO_{52}) from the Oxford Micro Beam Company [92] and yields a reduction coefficient 228 x 60 with about 7m length (from an object collimator to a target). The beam scanning is realized through a ferromagnetic scanning system with a scanning raster 0.5 x 0.5 mm² at about 10 Hz frequency. An electrostatic system of about 1MHz frequency and 50 x 50 µm² scanning raster was applied to accelerate a scanning process. A three-coordinate system (*XYZ*) of the Burleigh Inchworn Company allows a target positioning within 25 x 25 mm² and with 20 nm accuracy under a closed cycle operation. Using this micro-probe device, they obtained a spot area 290 x 450 nm² under 50nA proton beam current and 35 x 75 nm² H^{+2} ion beam area under about 1φA current.

In spite of an essential progress in parameters of the nuclear micro-probe devices and a number of advantages characterizing the focused ion beam of a mega-electron-volt energy in comparison with an ordinary electron beam and heavy ion beam of a low energiy, a future progress in a fabrication of commercial devices is related to a development of new systems for a micro-probe packaging allowing more compact devices. The authors of [93, 94] propose one version of such construction, which allows a reduction of device dimensions by a factor of five (to five meter length) and an improvement of the beam parameters at a target. Existing now probe-forming systems demonstrate very low reduction factor in comparison with *SEM* devices. Therefore, one needs a new probe system with about 1000 to 10000 reduction factor, which would be based on principally new lenses allowing the reduction of a working distance to 5mm. A system with such reduction factor would feature a high aberration. Therefore, a new type of an ion source of an essentially higher brightness is necessary. The authors of [89] report about a field source of He$^+$ ions featuring 10^9A/m^{2}cp brightness, which exceeds that of already existing high-frequency ion sources by seven orders of a magnitude. However, since a summary current of this source (about 100nA with an energy spread < 1 eV) is not high, a principally new scheme, which could stabilize a high voltage of an electrostatic accelerator, is necessary.

A metrological aspect of the focused electron beam device allowing a determination of the beam area at a target is well optimized. The authors of [88-91] consider one of the methods, which is employed to analyze a signal coming from a secondary ion when a special template is scanned and allows a determination of the beam area size at a level of several nanometers. Since a yield of secondary electrons is essentially lower when a light ion interacts with a matter and its transition inside a matter differs principally from an electron transition, a problem of standards, which could be employed for a determination of the beam area in < 30nm range, is crucial [90].

6.7.3. Interaction of Accelerated Charged Particles with Resistive Materials

When an accelerated charged particle passes through a sample, it gives rise to a change of physical and chemical properties of an irradiated region. In some cases, the irradiation region happens to be removed, when a sample is treated by a corresponding developer and an electro-chemical etching is used. In this case, a resistive material is called positive. When a non-irradiated surface is removed the resistive material is called negative. When a proton passes through a polymer material like the *PMMA*, a polymer chain breaks. As a result, the irradiated region contains molecular compounds of a low molecular weight. The irradiated region can be removed by a water solution of an isopropyl alcohol (*IPA*) 3 : 7 ratio (Fig. **31a**).

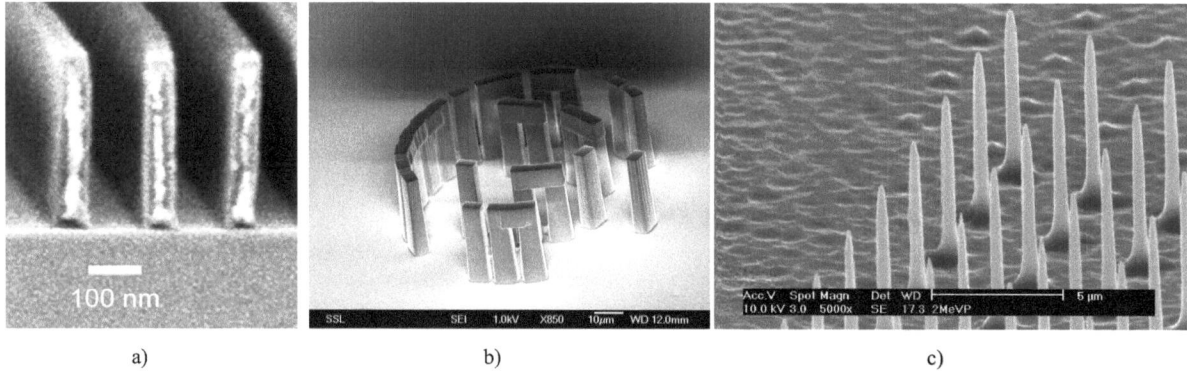

a)   b)   c)

Figure 31: An image of three-dimensional nanostructures formed with the help of p-beam writing [18]: *a*-a *SEM* image of parallel lines obtained in a *PMMA* layer of 350 nm thickness; *b*-a microscopic copy of a Stonehenge monumental construction formed in SU-8 single-layer material by variations of a proton energy; *c*-an array of needles with a high height-to-width ratio are formed in a bulky silicon volume (point curvature 15nm).

On the other hand, protons passing through SU-8 material form intermolecular transverse bonds, which increase a resistance to a chemical developer removing the non-irradiated region. Therefore, such resistive material is negative in reference to an irradiation process, which is performed by a charged particle beam (Fig. **31b**) [89, 90].

Radiation-induced defects of p-type silicon increase a specific resistance of the irradiated region. The electrochemical etching allows a formation of a porous structure in the non-irradiated region. Then, the porous structure is removed with the help of a potassium hydroxide solution. The technology allows a formation of three-dimensional nanostructures in a bulky silicon volume (Fig. **31c**). Similar mechanism is applied for a formation of three-dimensional structures in GaAs [93].

The main feature differing the proton beam technology from the electron beam one is that in the case of lighter electrons, a decreased structure size requires a lower thickness of the resistive material (Fig. **32**) (as a rule, the resistive material is deposited to a substrate by a centrifuge).

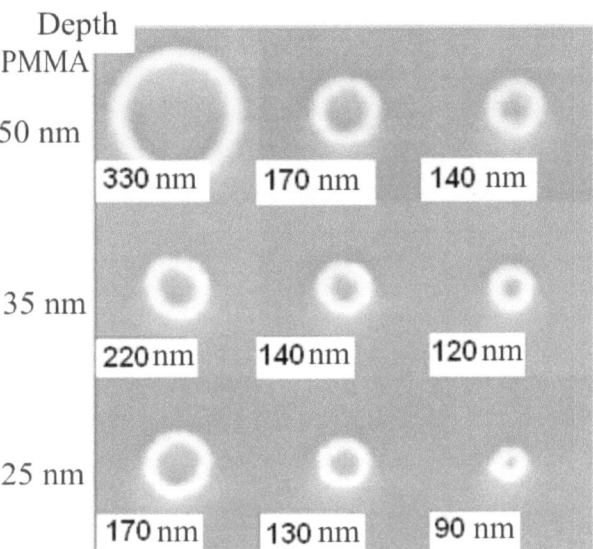

Figure 32: A dependence of a resistive material layer thickness (*PMMA*) on a surface nanostructure size formed using an *EBL* technology.

This is because of an electron scattering, which occurs when the particle passes through a sample. In this case the structures are practically two-dimensional. At the same time, the proton beam is not broadened in the resistive material, allowing a formation of three-dimensional nanostructures. In order to have a desired result, a dose value suitable for the resistive material should be taken into account. In the case of the proton beam, according to evaluations of [93], it is lower by two orders of a magnitude than it is necessary for the electron beam. Table **3** presents doses for various materials [93, 94].

Table 3: An application of various resistive materials in an *EBL* technology.

Resistive Material	Type	Necessary Dose nK/mm^2	Smallest Resulting Characteristic Dimension
PMMA	positive	80-150	20-30nm
SU-8	negative	30	60nm
HSQ	negative	30	22nm
PMGI	positive	150	1.5μm
WL-7154	negative	4	800nm
TiO$_2$	negative	8000	5μm
Si	negative	80 000	15nm (a needle point)
Dial Plate	negative	10	10μm
ADEPR	negative	125-238	5μm
Forturan	positive	1	3μm
PADC (CR-39)	positive	600	5μm
Ma-N 440	negative	200	400nm
GaAs	negative	100 000	12μm

QUESTIONS FOR CONTROL

1. Enumerate main methods employed for a nanomaterial fabrication.

2. What are features of a nanopowder fabrication?

3. What is compacting?

4. What are the ways of a fabrication of a nanocrystalline material without pores?

5. What are the disadvantages of an intensive plastic deformation method employed for a fabrication of a nanocrystalline material?

6. What methods form a basis of a thin-film technology employed for a fabrication of a nanostructured film and a coating?

7. What is a difference between a magnetron sputtering and a vacuum deposition?

8. What methods are employed for a fullerene fabrication?

9. What atomic substances are efficient catalysts of a CNT growing, and, in particular, for a formation of an oriented CNT array?

10. What are the basic features characterizing a CNT growing using an arc-discharge method?

11. What is a role of a template in a formation of an oriented CNT array?

12. What are the features of an accelerated low-energy electron, ion, and light ion beams of an average energy passing in a matter? What is a difference of their passing ability?

13. What is a difference between an electron beam lithography and a scanning electron microscopy?

14. What is a difference between a proton beam in comparison with an electron beam application for a resistive material?

REFERENCES

[1] Gusev AI, Rempel AL. Nanocrystalline Materials. Moscow: Physmatlit 2001.
[2] Pozdniakov VA. Physical Material Science of Nanostructured Materials. Moscow: MGIU 2007.
[3] Suzdalev IP. Physical Chemistry of Nanoclusters, Nanostructures, and Nanomaterials. Moscow: Komkniga 2006.
[4] Alymov MI, Zelenskii VA. Methods of Formation and Physical-Mechanical Properties of Volume Nanocrystalline Materials. Moscow: MIFI 2005.
[5] New Materials. Ed. By Karabasov YuS. Moscow: MISIS 2002.
[6] The New in FabricationTechnology of Materials. Ed. By Osipian Yu.A and Hauff A. Moscow: Mashinostroenie 1990.
[7] Gunter B, Kumpmann A. Ultrafine Oxide Powders Prepared by Inert Gas Evaporation. Nanostruct. Mater 1992; 1: 27-30.
[8] Kotov YuA, Yavorskii NA. Investigation of Particles Formed under Electrical Explosion of Semiconductors. Fiz Khim Obrab Mater 1978; 4: 24-30.
[9] Ivanov VV, Kotov YA, Samatov OH, *et al*. Synthesis and Dynamic Compaction of Ceramic Nanopowders by Techniques Based on Electric Pulsed Power. Nanostruct Mater 1995; 6: 287-290.
[10] Gen MYa, Miller AV. Method of Formation of Ultradispersion Metallic Powders. Poverkhnost. Fizika Khimiia Mekhanika 1983; 2: 150-154.
[11] Champion Y, Bigot J. Preparation and Characterization of Nanocrystalline Copper Powders. Scr Met 1996; 35: 517-522.

[12] Blagoveschenskii YuV, Panfilov SA. Jet-Plasma Processes for Powder Metallurgy. Elektrometallurgiia 1999; 3: 33-41.
[13] Kiparisov SS, Padalko OV. Equipment for Plants of Powder Metallurgy. Moscow: Metallurgiia 1988.
[14] Karlov NV, Kirichenko MA, Lukianchiuk BS. Macroscopic Kinetics of Thermal-Chemical Processes under Laser Heating: State of the Art and Perspectives. Russ Chem Rev 1993; 62: 223-243.
[15] Powder Metallurgy. Materials, Technology, Properties, Application Fields. Ed. By Fedorchenko IM. Kiev: Naukova Dumka 1985.
[16] Leontiev ON, Alymov MI, Teplov OA. Hetero-Phase Synthesis of Iron-Copper Powders. Fiz Khim Obrab Mater 1996; 5: 105-109.
[17] Kriechbaum GW, Kleinschmidt P. Superfine Oxide Powders-Flame Hydrolysis and Hydrothermal Synthesis. Angew Chem Adv Mater 1989; 101: 1446-1453.
[18] Bykov Y, Gusev S, Eremeev A, *et al.* Sintering of Nanophase Oxide Ceramics by Using Millimeter-wave Radiation. Nanostruct Mater 1995; 6: 855-858.
[19] Chen I-W, Wang XH. Sintering Dense Nanocrystalline Ceramics Without Final-Stage Grain Growth. Nature 1996; 404: 168-171.
[20] Alymov MI, Leontieva ON. Synthesis of Nanoscale Ni and Fe Powders and Properties of Their Compacts. Nanostruct Mater 1995; 6: 393-395.
[21] Kovernisty YuK. Volume-Amorphisation of Metallic Alloys. Moscow: Nauka 1999.
[22] Sudzuki K, Fujimori H, Hasimoto K. Amorphous Metals. Moscow: Metallurgiia 1987.
[23] Zolotukhin IV. Physical Properties of Amorphous Metallic Materials. Moscow: Metallurgiia 1986.
[24] Valiev RZ, Aleksandrov IV. Nanostructured Materials Formed by Intensive Plastic Deformation. Moscow: Logos 2000.
[25] Bunshah RF, *et al.* Deposition Technologies for Films and Coatings. Park Ridge, New Jersey (USA): Noyes Publications 1982.
[26] Nikitin MM. Technology and Equipment for Vacuum Deposition. Moscow: Metallurgiia 1992.
[27] Technology for Thin Films. Ed. By Maissel L. and Gleng R. Moscow: Sov. Radio. 1970; 1: and 2.
[28] Komnik YuF. Physics of Metallic Films. Moscow: Atomizdat 1979.
[29] Andreev AA, Sablev VP, Shulaev VM, Grigoriev SN. Vacuum-Arc Devices and Coatings. Kharkov: NNTs "KhFTI" 2005.
[30] Malik A, Raunt RI. eds. New Nanotechniques. Chapter 2 "Structure and Properties of Protective Composite Coatings and Modified Surfaces Prior and After Plasma High Energy Jets Treatment. In: Pogrebnjak AD, Shpak VM, Beresnev VM. Nova Science Publ 2009; 4: 25-114.
[31] Danillin BS, Syrchin VK. Magnetron Sputtering Systems. Moscow: Radio I Sviaz 1982.
[32] Kostrezhetskii AI, *et al.* Reference Book for Operators for Coating Deposition in Vacuum. Moscow: Mashinotsroenie 1991.
[33] Makarova TP. Electrical and Optical Properties of Mono-dimensional and Polymerizated Fullerenes. Fiz Tekh Poluprovodn 2001; 35: 257-293.
[34] Naschiokin AV, Kolmakov AG, Soshnikov IP. Application of Multifractal Concept for Characterization of Properties of Composite Films of C60 Fullerenes Doped by CdTe. Tech Phy Lett 2003; 29: 8-14.
[35] Caricato AP, Barucca G, Di Cristoforo A, *et al.* Excimer Pulsed Laser Deposition and Annealing of YSZ Nanometric Films at Si Substrates. Appl Surf Sci 2005; 248: 270-275.
[36] Kobea S, Zuzeka K, Sarantopoupou E, *et al.* Nanocrystalline Sm-Fe Composites Fabricated by Pulsed Laser Deposition at 157nm. Appl Surf Sci 2005; 248: 349-354.
[37] Amoruso S, Ausanio G, De Lisio C, *et al.* Synthesis of Nickel Nanoparticles and Nanoparticles Magnetic Films by Femtosecond Laser Ablation in Vacuum. Appl Surf Sci 2005; 247: 71-75.
[38] Belyi AV, Karpenko GD, Myshkin NK. Structure and methods of Formation of Wear Resistant Surface Layers. Moscow: Mashinostroenie 1991.
[39] Belianin AV, Krivchenko VA, Lopaev DV. Nanostructured ZnO Films for Devices of Microelectronics and Optics. Teckhnologiia I Konstruirovanie v Elektronnoi Apparature 2006; 6: 48-54.
[40] Shulaev VM, Andreev AA. Superhard Nanostructured Coatings in NNTs KHFTI. Physical Surface Engineering 2008; 6: 4-19.
[41] Tolok VT, Shvets OM, Lymar VF, *et al.* 1757249 Russia, MKI C23 C14/00. N 4824783/SU.
[42] Azarenkov NA, Beresnev VM, Pogrebnjak AD. Structure and Properties of Protecvite Coatings and Modified Layers of Materials. Kharkov: KhNU 2007; 565.
[43] Jagodkin YuD. Ion-Beam Treatment of Metals and Alloys. Itoghi Nauki I Tekhniki. Metallovedenie I Termicheskaia Obrabotka Metallov. Moscow: VINITI 1980; 14: 142-185.
[44] Abrojan IA, Andronov AN, Titov FT. Physical Fundamentals of Electron and Ion Tehcnology. Moscow: Vysshaia Shkola 1984.
[45] Zolotyukhin IV, Kalinin YuE, Stognei OV. New Directions of Physical Material Science. Voronezh: VGU 2000.
[46] Kadyrzhanov KK, Komarov FF, Pogrebnjak AD, *et al.* Ion-Beam and Ion-Plasma Modification of Materials. Moscow: MGU 2005; 640.

[47]   Poate JM, Foti G, Jacobson DS. Surface Modification and Alloying by Laser, Ion, and Electron Beams. New York: Plenum Press 1983.
[48]   Redi J. Industrial Application of Lasers. Moscow: MIR 1981.
[49]   Cheng Li, Plot K. Molecular-Beam Epitaxy. Moscow: MIR 1989.
[50]   Syrkin VG. CVD Method-Chemical Vapor-Phase Deposition. Moscow: Nauka 2000.
[51]   Andrievskii RA. Formation and Properties of Nanocrystalline Refractory Compounds. Russ Chem Rev 1994; 63: 431-448.
[52]   Semkina TV, Vengher EF. Physical-Chemical Fundamentals of Formation and Modification of Micro-and Nanostructures. FMMN 2008; 1: 76-79.
[53]   Lozovik YuE, Popov AV. Formation and Growth of Carbon Nanostructures-Fullerenes, Nanoparticles, Nanotubes, and Cones.Physics-Uspekhi 1997; 167: 751-774.
[54]   Nerushev OA, Sukhinin GI. Kinetics of Formation of Fullerenes by Electro-Arc Evaporation of Graphite. Tech Phys 1997; 67: 41-49.
[55]   Gorelik OP, Diuzhev GA, Novikov DV, *et al.* Cluster Structure of Particles of Fullerene-Containing Black and Fullerene C60 Powder. Technical Physics 2000; 70: 118-125.
[56]   Eletskii AV. Carbon Nanotubes and Emission Properties.Physics-Uspekhi 2002; 172: 401-438.
[57]   Diachkov PN. Carbon Nanotubes: Structure, Properties, Application. Moscow: BINOM 2006.
[58]   Tkachiov AG, Zolotukhin IV. Equipment and Methods for Synthesis of Solid Nanostructures. Moscow: Mashinostroenie 2007.
[59]   Komarov FF, Mironov AM. Carbon Nanotubes-the Present and Future. Phys Chem Solids 2004; 5: 411-429.
[60]   Ressier L, Grisolia J, Martin C, *et al.* Fabrication on Planar Cobalt Electrodes Separated by a Sub-10nm Gap Using High Resolution Electron beam Lithography with Negative PMMA. Ultramicroscopy 2007; 107: 985-988.
[61]   Lee RA, Patrik W. Optical Image Formation Using Surface Relief Micrographic Picture Elements. Microelectronic Engineering 2007; 84: 669-672.
[62]   Gierak J, Septier A, Vien A. Design and Realization of a Very High-Resolution FIB Nanofabrication Instrument. Nucl Instr Meth 1999; A427: 91-98.
[63]   Chyr I, Steck AJ. GaAs Focused Ion Beam Micromachining with Gas-Assisted Etching. J Vac Sci Technol 2001; B19: 2547-2550.
[64]   Watt F, Bettiol AA, Van Kan JA, *et al.* Ion Beam Lithography and Nanofabrication: A Review. International Journal of Nanoscience. 2005; 4: 269-286.
[65]   Watt F, Breese MB, Bettiol A, Van Kan JA. Proton Beam Writing. Mater Today 2007; 10: 20-29.
[66]   Hovington P., Drouin D., Gauvin R. *et al.* A new Monte Carlo code inc language for electron beam interactions-part III stopping power at low energies. Scanning 1997; 19: 29-35.
[67]   http://www.gel.usherb.ca/casino/index.html.
[68]   Yamashita T. PhD Thesis, Georgia Institute of Technology, august 2005.
[69]   http://www.jeol.com.
[70]   Reyntjens S., Puers R. A review of focused ion beam applications in microsystems technology. J. Micromech. Microeng 2001; 11: 287-300.
[71]   Watt F, Van Kan JA, Rajta I, *et al.* The National Nniversity of Singapore high ion nano-probe facility: Performers test. Nucl Instr Meth 2003; B210: 14-20.
[72]   Jeroen A, Van Kan JA, Bettiol AA, *et al.* Proton beam writing: a progress review. Int J Nanotech 2004; 1: 464-477.
[73]   Mous DJ, Haitsma RG, Butz T, *et al.* The novel ultrastable HVEE 3,5 Mv singletron accelerator for nanoprobe application. Nucl Instr Meth 1997; B130: 31-36.
[74]   Breese MB, Grime GW, Linford W, *et al.* An extended magnetic quadrupole lens for, a high-resolution nuclear. Nucl Instr Meth 1999; B158: 48-57.
[75]   Ignat'ev IG, Magilin DV, Miroshnichenko VI, Ponomarev AG, *et al.* Immersion probe-forming system as a way to the compact design of nuclear microbe. Nucl Instr Meth 2004; B231: 94-100.
[76]   Storigko VYu, Ponomarev OG, Moroshnichenko VI. UA 67341A, G01N23/00, 2003038121. 2004.
[77]   Tondare VN. Quest for brigthess, monochromatic noble gas ion sources. J Vac Sci Technol 2005; A23: 1498-1508.
[78]   Morgan J, Notte J, Hill R, *et al.* An Introduction to the Helium ion microscope. Microscopy Today 2006; 14: 24-31.
[79]   Watt F, Rajta I, Van Kan JA, *et al.* Proton beam micromachined resolution for nuclear microprobes. Nucl Instr Meth 2002; B190: 306-311.
[80]   Van Kan JA, Sunchez JL, Xu B, *et al.* Resist materials for proton micromachining. Nucl Instr Meth 1999; B158: 179-184.
[81]   Mistry P, Gomez-Morilla I, Grime GW, *et al.* New developments in the applications of proton beam writing. Nucl Instr Meth 2005; B237: 188-192.
[82]   Van Kan JA, Bettiol AA, Chiam SY, *et al.* New resists for proton beam writing. Nucl Instr Meth 2007; B260: 460-469.
[83]   Aliofkhazraei M. Nanocoatings: Size Effect in Nanostructured Films Springer 2011.
[84]   Decher G, Schlenoff JB. Multilayer Thin Films. Sequential Assembly of Nanocomposite Materials. Berlin: Wiley-VCH Verlag 2002.
[85]   Masuda Y. Nanocrystals. SCIYO 2010.

[86] Ramsden J. Essentials of Nanotechnology. Ventus Publishing ApS 2009.
[87] Bhushan B. Scanning Probe Microscopy in Nanoscience and Nanotechnology Springer 2010.
[88] Gopalakrishnan S, Mitra M. Wavelet Methods for Dynamical Problems: With Application to Metallic, Composite, Nano-Composite Structures. CRC Press 2010.
[89] Sattler KD. Handbook of Nanophysics: Nanoelectronics and Nanophotonics. CRC Press 2010.
[90] Gross R, Sidorenko A, Tagirov L. Nanoscale Devices-Fundamentals and Applications. NATO Science Series II: Mathematics, Physics and Chemistry, Springer 2006; 233.
[91] Books New Nanotechnologies/edit. A. Malik and R.J. Rawat. A.D.Pogrebnjak, A.P.Shpak, V.M.Beresnev. Chapter 2 (p. 25-114). Structure and Properties of Protective Composite Coatings and Modified Surface Prior and After Plasma High Energy Jets, Nova Science Publisher. 2009 : 687.
[92] Pogrebnjak AD, Lozovan AA, Kirik GV, *et al.* Structure and Properties of nanocomposite, hybrid and polymers coatings,Publ. House URSS, Moscow, 2011, 344.
[93] Azarenkov NA, Beresnev VM, Pogrebnjak AD, *et al.* Fundamentals of Fabricated Nanostructured Coatings, Nanomaterials and Their Properties, Publ. House URSS, Moscow 2012; 352.
[94] Pogrebnjak AD, Shpak AP, Beresnev VM, *et al.* "Structure and Properties of Nano-and Microcomposite Coating Based on Ti-Si-N/WC-Co-Cr" Acta Physica Polonica A, 2011; 120, No. 1: 100-104.

CHAPTER 7

Methods of Nanomaterials Investigation

Abstract: This Chapter briefly describes the methods, which are employed to study a nano-material (structural and chemical analysis). Such methods as AFM (an atomic force microscopy), STM (a scanning tunneling microscopy), XRD, SIMS, mechanical tests, which are employed to study a nano-material (measurements of a nanohardness and an elastic modulus), are considered.

Keywords: TEM, STM, AFM, SIMS, RBS, nanohardness.

7.1. INTRODUCTION

A modern development of physics and technology dealing with a solid nanostructure exhibiting a continuous transition of a structure element topology from a submicron size to a nanosize geometry requires the new improved diagnostic methods and the new equipment to analyze properties and processes occurring in a small-grain system, a nanomaterial, and an artificially formed nanostructure. In this sense, a special attention is paid to a development and an application of high resolution methods employed for a practical diagnostics and a nanostructure characterization, which could supplement each other and provide the most complete information about basic physical, physical-chemical, and geometrical parameters of the nanostructure and the processes, which take place in this nanostructure [1-3].

Today, there is a great number of diagnostic methods, a huge number of methods employed to study the physical, physical-chemical parameters, and the characteristics of a solid and a molecular structure. At the same time, a formation of a nanostructure, a small-grain system, and a nanomaterial with a desired property, which could be applied in a modern engineering, sets new tasks of an employed diagnostics. To solve the modern problems of the nanostructure diagnostics, an adaptation of the traditional methods to these tasks and a development of new local methods (to 0.1nm range), which could be employed to study and analyze the properties and processes characterizing a nanogeometry and a small-grain system, is necessary.

The nanodiagnostics must be nondestructive and yield the information both about a nanostructure and about an electron property with an accuracy of ranging at an atomic level. A crucial factor in the development of nanotechnologies is also a possibility to control an atomic and an electron process with a high time resolution accuracy. An ideal time is equal or shorter than an atomic oscillation period (to 10^{-13}s and shorter). An electronic, an optical, a magnetic, a mechanical, or other property of nanoobjects also need studies at a "nanoscopic" level.

The following methods of structural and chemical analysis, which take into account a specific character of the nanomaterial, are employed-an electron microscopy, a diffraction, and a spectral analysis.

7.2. METHODS OF STRUCTURAL AND CHEMICAL ANALYSES OF NANOOBJECTS

An electron microscopy. A method serves to obtain the information about a material nanostructure (a particle size), to oversee an interface, *i.e.* to study a structure of a matter or a material. The operation principles of various microscopes are strongly different. Some of them are based on an electron transmission through a sample (a transmission electron microscopy), others employ an electron reflection from a sample (a reflection electron microscopy, a slow electron microscopy, and a scanning electron microscopy), an ion reflection (a field ion microscopy), a surface scanning by an electron beam (a scanning electron microscopy), or probing needles (a scanning electron microscopy, an atomic force microscopy). The majority of microscopy methods can provide a nanorange resolution. The field ion, a scanning tunneling, and the atomic force microscopy yield a microscopic image, an accuracy of which is of an atomic range [1-10].

Alexander D. Pogrebnjak and Vyacheslav M. Beresnev
All rights reserved-© 2012 Bentham Science Publishers

A transmission electron microscopy yields a high-resolution image and a microdiffraction pattern of the same region for one experiment. A modern transmission electron microscope is able to provide up to 0.1 nm resolution from up to 50 nm area. An obtained image yields the information about a material structure, and a diffraction pattern shows a crystal lattice type. This method can be applied to analyze the microdiffraction. A diagram for an area under study looks like points depending on a material content (a singlecrystal or a polycrystal with a grain size exceeding the studied area), continuous, or discrete reflexes (a very small crystallite in a grain or several small grains). An orientation of crystals and a disorientation of grains and subgrains can be determined using the microdiffraction analysis. Fig. **1** demonstrates an electron-microscopy profile image for a cobalt particle on Al_2O_3 substrate [3].

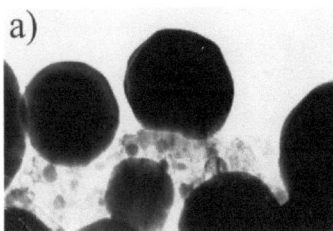

Figure 1: An electron-microscopy image of a changing Co particle profile on Al_2O_3 substrate: *a*-above the boundary temperature T_b; *b*-below the boundary temperature T_b.

A very narrow beam of a transmission electron microscope is employed for a local chemical analysis of materials and to analyze light elements (boron, carbon, oxygen, nitrogen) using a spectrum of electron energy loss after their pass through a studied material [1, 3, 4].

An ion-field microscopy. This method allows a resolution, which is approximately equal to an interatomic-range. A positive potential is applied to a metallic needle with a sharp tip, which is placed in a chamber under high vacuum. Both an electric field and its gradient in the tip vicinity are so high, that the residual gas molecules are ionized coming closer to the tip vicinity, transfer electrons to the needle, and gain a positive charge. These gaseous cations (basic ions) are repulsed by the needle and fly from it along the electrostatic field lines to a closely positioned photo-plate. Then they fall to the plate and form the luminescent points. Every point corresponds to one atom from the probe tip. A point distribution over the plate looks like a strongly magnified image of the atomic distribution at the needle tip.

A scanning electron microscopy (SEM). An efficient way to obtain an image of a sample surfaces is a surface scanning by an electron beam. A resulting raster is similar to that formed by an electron gun scanning employed in a TV-screen. The information about the surface may be obtained also with the help of a solid-state probe, which scans some individual surface regions. Scanning may be realized also by a probe, which can measure a current formed by an electron tunneling between a sample surface and the probe tip, or by a probe, which can measure a force of the surface and needle tip interaction. A basic function of the scanning electron microscopy is a visualization of a topography and map of an element distribution in a surface. Fig. **2** shows a scanning electron microscopy microphotograph for Au film surface, which is deposited to TiO_2 substrate (110) and annealed at 550°C [7-9].

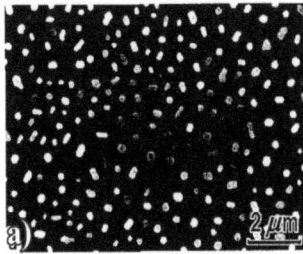
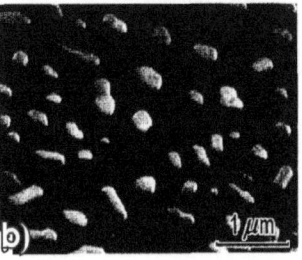

Figure 2: Images of Au film of 120 Å thickness obtained by a scanning electron microscopy: *a*-for a normally beam incidence; *b*-under 45 angle beam incidence [9].

A scanning tunneling microscopy (STM). A scanning tunneling microscope uses a needle with a superthin tip as a probe. This tip is switched to a positive pole of a voltage source and is brought closer to a studied surface-to about 1nm distance. The electrons relating to a concrete atom of a sample surface are attracted to a positively charged tip, jump (tunnel) to it, and generate a weak electric current. A need to employ, in any way, a conductive material, to sustain a superhigh vacuum and a low temperature (up to 50-100 K) for a high resolution is considered as a disadvantage of this method. At the same time, these requirements are not necessary for about 1nm-range resolution.

Fig. 3 shows a photo of a quasihexagonal reconstructed surface of Pt(100) taken from [8]. An elementary cell with a superstructure contains more than 30 atoms of [0$\overline{1}$1] orientation and six atoms of [011].

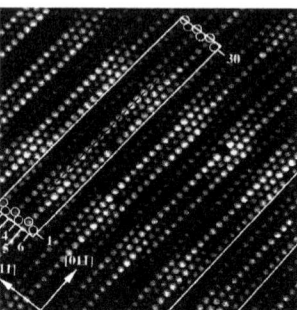

Figure 3: A scanning tunneling microscopic image for a quasihexagonal reconstructed surface of Pt (100) [9].

An atomic-force microscopy (AFM). This method records a change of a probe tip (needle) interaction force with a studied surface. The needle is located at the end of a cantilever beam of a known stiffness. The cantilever beam is able to bend under an action of a weak force arising between the sample surface and the tip. The force may be Van der Waals (molecular), electrostatic or magnetic depending on the method variation. The cantilever deformation is measured by deviation of a laser beam falling to its back surface or by a piesoresistive effect arising in a material when the cantilever is bended [9].

A fundamental difference between the scanning tunneling and the atomic force microscope is that the first one measures a tunneling current between the probe and the surface, and the second one measures an interaction force between them. Similar to scanning tunneling microscope, the atomic force microscope has two operation regimes. The atomic force microscope can function in a contact with a surface, when a main role is played by a repulsion force of the electron shells of probe and surface atoms, and without a contact, when the probe is far from the sample and the Van der Waals force is dominating. Similar to the scanning tunneling microscope, it employs a piesoelectrical scanner. In the process of scanning, a probe vertical motion may be controlled by a change of an interference picture, which is imaged by a light beam directed by an optical fiber. Fig. **4** shows an image of a surface morphology for Ti-Al-N ion-plasma deposited coating obtained using the atomic-force microscope (*AFM*). T_{ip} radius is ~ 10 nm. Measurements are performed in air [10-14].

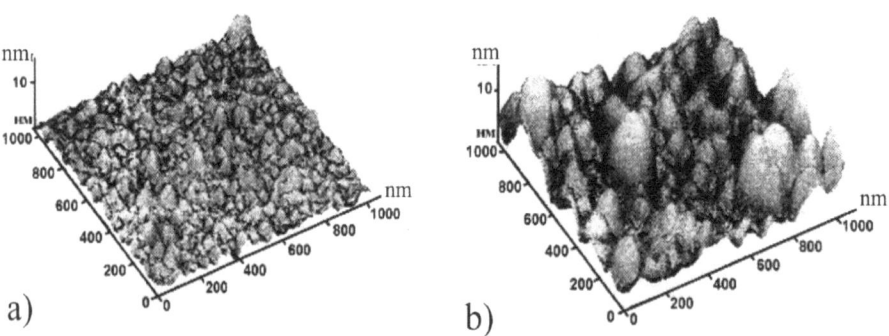

Figure 4: Topographic image of Ti-Al-N coating obtained using *AFM*: a-U_{cm}= 100V; b-U_{cm} = 200 V [10].

An obvious advantage of the atomic force microscopy is that it can be applied to study any type of a surface of a conducting, a semiconducting, and a dielectric material. Modern devices allow measurements of a needle friction power, mapping of an elasticity of a studied material region, tests of a wear resistance using scratching. The semiconducting diamond needles are employed to measure a sample surface capacity, a conductivity of a near surface layer, and to determine an impurity concentration using a change of the capacity value. A plane resolution (x-and y-coordinate) is about 1nm, and a height one (z-coordinate) is up to 0.1 nm. A limitation of this method is a needle material hardness. However, for most materials under study the diamond or a fullerite needle is good.

All three described scanning microscopes yield the information about a topography and surface structure defects with a resolution accuracy, which is close to an atomic range.

An X-Ray analysis. This method allows a qualitative phase analysis, a determination of a lattice parameter, an atomic displacement, a calculation of coherent scattering areas, a value of microdistortions with a high accuracy [5]. If an X-ray pattern shows a phase, a structure of which cannot be determined by these methods, every phase can be identified individually by comparing diffraction angles Θ (or interplanar distances d_{HKL}) with the data for this phase, which are expected in this sample according to the results of an element analysis and a phase diagram. In this case, the reference tables for the interplanar distance and the relative line intensities, as well as the computer data base *PCDFWIN* are employed. The References (tables) contain a set of values for the interplanar distance and the intensity, which is arranged in the order of a decreasing d value. The References indicate three d values for the planes with a strong reflex (the ordering succession corresponds to a decrease of the reflex intensity). The fourth value corresponds to the greatest interplanar distance for this matter. In addition, the references indicate a name, a matter chemical formula, the parameters of an elementary cell, a crystalline system, a space group, and some physical characteristics. The *ASTM* references contain more than 25 thousand of etalon spectra and 1500 to 2000 etalons are annually added to it.

The diffraction method allows a calculation of a grain size averaged over a studied matter volume, while the electron microscopy is a local method and can determine an object size only in a limited observation field. To determine an *RCS* (a region of a coherent scattering) value, various methods are employed-an approximation method, which can approximately evaluate a true diffraction broadening, a Stock's method allowing a selection of a diffraction broadening curve without any assumptions like a function describing a line profile, and a method of a harmonic (Fourier) analysis. The harmonic analysis determine a size of up to 10 to 15nm blocks and a microdistortion exceeding 4×10^{-4}.

Fig. 5 shows X-ray patterns for a metallic nickel with about 2 to 10μm grain size and for a compacted nanocrystalline nickel with about 20 nm grain, which were taken from the work [8]. Ni transition to a nanostate results in an essential broadening of the diffraction lines.

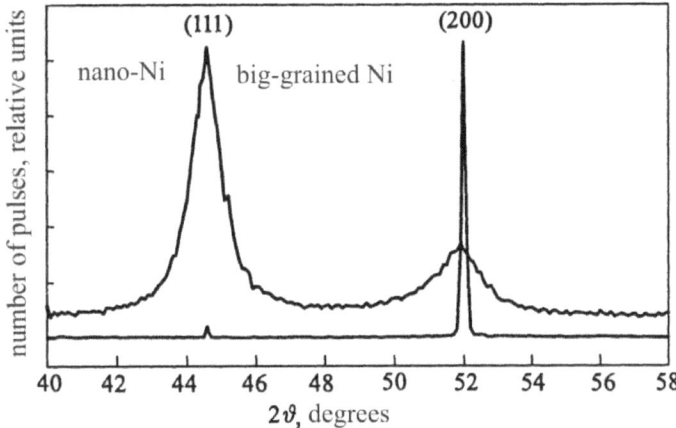

Figure 5: A comparison of X-ray patterns for a big grained and a compacted nanocrystalline nickel [8].

In general case, the X-ray diffraction is not the method, which is able to yield the information about a surface structure, since a surface scattering is five orders of a magnitude weaker than a volume one. In this connection, the data about a structure can be obtained from a sliding incidence of X-ray emission, when the incidence angle is equal or lower than a critical angle for a complete inside reflection.

Methods of spectral analysis. Methods, which are employed to study a solid surface, are based on an analysis of the energy spectra of a reflected emission induced in a material under study by an electron irradiation. Today, several scores of such methods are known. However, not all of these methods have a universal or a concrete application for the nanomaterial. For example, a quantitative ability of a widely known X-ray spectral microanalysis is not better than 1 to 2 μm in diameter; an ability of an X-ray photo-electron microscopy is 2 to 10 mm. In this connection, below we shall consider a number of methods, an ability of which, on one hand, is interesting for nanomaterial researches, on the other hand, which are most indicative and popular [9-13].

A Secondary ion-mass spectrometry method (SIMS). This method is one of the physical methods, which is employed to investigate a surface. It is employed to have the data concerning a quantitative distribution of impurities over a surface layer depth in different materials. The analysis is performed under a high vacuum. A sample surface is bombarded by a primary ion beam of 0.1 to 100 keV energy. Falling to the surface, the primary ions knock out secondary particles, a fraction of which (usually 5 %) in an ionized state leaves the surface. These ions are focused and move to a mass-analyzer, in which they are separated accroding to their mass and charge. Then, they arrive to a detector, which fixes a secondary ion current intensity and yields the information to a computer. Fig. **6** shows a mass spectrum of a high temperature superconducting material $YBa_2Cu_3O_7$.

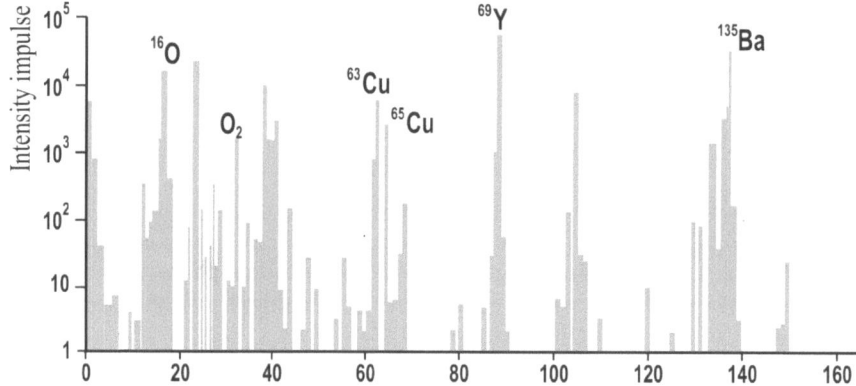

Figure 6: A spectrum of secondary ion clusters of a high-temperature superconductor $YBa_2Cu_3O^7$ (a primary ion energy O_2^+ is 4 keV) [13].

A range of the peak intensities, which are seen in this spectrum is typical: is covers more than five orders of a magnitude. Not many methods have such wide dynamical range. This wide range and a high sensitivity are the advantages of this method. A presence of molecular ions among those, which are sputtered from a matter surface, makes this method especially valid for studies of a molecular cluster and molecular adsorber formations: the *SIMS* spectrum has a characteristic form, which is definitely related to a molecule sort.

An electron Auger spectroscopy (AES). An *AES* method is one of the most popular spectroscopic methods, which is applied to analyze a chemical composition of a sample near-surface layer. The layer thickness, which can be determined by the *AES* method, corresponds to an average free path of Auger-electrons. A depth varies from about 0.5 nm (for 50 eV energy) to about 2 nm (for 500 eV energy). In such a way, a spectrum for a low energy region, is most convenient for a detection of surface particles. A minimum area of analysis is limited by an electron spot diameter and is 0.01 to 0.1mm. A sensitivity of the Auger method depends on an element, which is aimed to be detected. In practice, a characteristic peak may be detected, if a relative surface atomic concentration is 0.1 to 1 %. An application of a raster technique allows a two-dimensional surface analysis, and in a combination with an ion sputtering-a three-dimensional one.

A Rutherford back scattering spectroscopy. A Rutherford back-scattering spectroscopy (*RBS*) is one of the most popular methods, which is applied for a quantitative analysis of elements in a surface layer. It is applied for a very wide circle of materials and may serve an etalon for other methods of analysis. This method has a highest accuracy when it is applied to analyze an element, an atomic mass of which is higher than a matrix one. A main principle of the Rutherford back-scattering method (*RBS*) is that a sample surface is irradiated by an ion beam of 1 to 3 MeV energy (usually He^+ or H^+ ions). As a rule, the beam diameter is 10μm to 1mm. The *RBS* spectrum is a plot: an abscissa (x-coordinate) is the number of an energy channel (n_i), in which a scattered ion of a definite energy occurs and an ordinate (y-coordinate) is an amount of ions (H_i) occurring in the *ni* channel. Depending on an analyzer type (semiconducting, magnetic, *etc.*), one can obtain various characteristics of a back-scattered particle. A yield of this back scattering from a sample surface manifests itself as a surface peak. An analysis of its intensity gives the information about a surface structure. Fig. **7** shows an *RBS* spectrum of helium ions, which are scattered from Al-Ni film on a molybdenum substrate. This spectrum is a common type among the *RBS* spectra, which can be met in the experiments, which employ the Ruitherford back-scattering method.

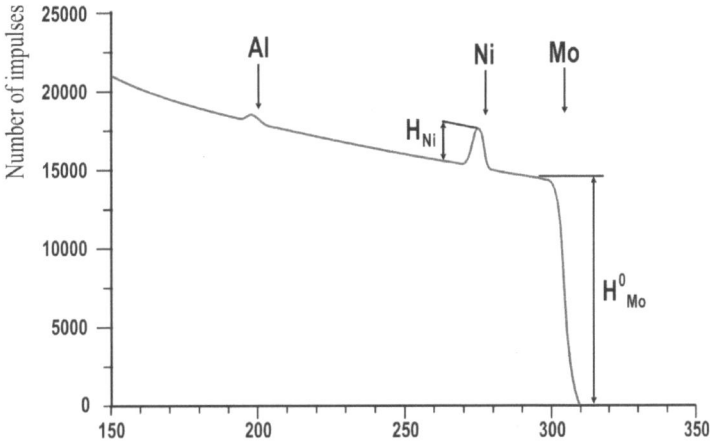

Figure 7: An RBS spectrum for Al-Ni film-molybdenum substrate system [13].

This method is employed to perform a qualitative element analysis of a sample, to study a surface composition, to perform a quantitative element analysis, to determine a film thickness, and to get knowledge of an element depth distribution (*i.e.* to plot a concentration profile) [22-28].

Taking into account specific peculiarities of a nanoobject and its characteristic size, a further development of various diagnostic methods (in particular, such diagnostics, which could be "incorporated" into a technological process) seems to become now a very important part of high technologies aimed to a fabrication and an analysis of properties of a new generation nanostructure.

7.3. TESTS OF MECHANICAL NANOHARDNESS OF SOLIDS

A surface structure and properties determine many servicing properties of a tool. Modern high technologies, which are now applied in engineering for modification of a surface layer, are able to form films and coatings with a unique combination of properties (including a nanosize objects). The obtained properties principally differ from the properties of a usual material, which is treated by traditional methods. An application of nanotechnologies into a modern electronic engineering requires studies of material physical, mechanical, and tribological properties, which could be realized at a submicron and a nanometer level [28-31].

Recently, a method of low load continuous indentation (a nanoindention) is employed to determine mechanical characteristics, such as a hardness and an elastic modulus of a surface layer. An indentation is perfored within a depth range from several tens to the hundred of a nanometer. The nanoindentation and a microindentation are employed to study a micromechanical behavior and a structure sensitivity of mechanical properties in a small sample, a thin film, and a coating [14-17]. Fig. **8** shows a nanoindenter. Mainly, the nanoindenter is equipped by a diamond Berkovich tripod pyramid.

A computer program sets the parameters for testing procedure: a load, a loading rate, an endure time, and a relief time.

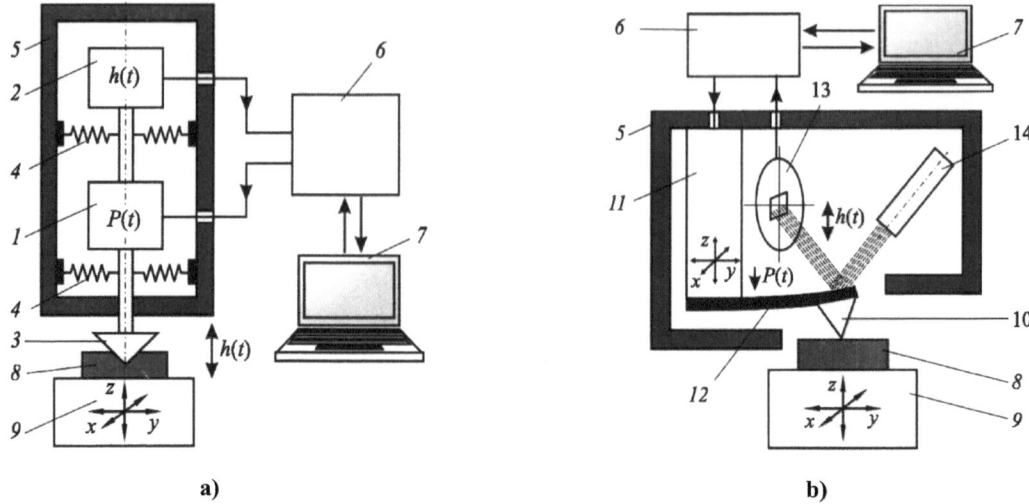

Figure 8: A nanoindenter (*a*) and an atomic force microscope (*b*): *1* a force cell; *2*-a sensor to control a movable rod with the indenter (*3*); *4*-rod hanger springs; *5*-a body of a measuring head; *6*-a controller block; *7*-a computer; *8*-a sample; *9*-a stage; *10*-a probe; *11*-a piezoelectric actuator; *12*-a consol microcantilever; *13*-a four-window photodetector (indicating the probe motion); *14*-a laser [15].

The device is composed of a loading site (*1*), a precision sensor (*2*) to indicate the indentation pyramid motion (*3*) with springs (*4*), which are joined in a single measuring head (*5*), a controller block (*6*), and a computer (*7*) with a software, to control all operation cycles of the device, an accumulation, a treatment, and a data storage. An optical microscope is employed to select an indentation point (*8*). A stage helps to position and move a sample along two or three coordinates (*9*). In more perfect devices, the stage is automatic with a computer control. A set of sites, their functions, and interactions of the nanoindenter and the atomic force microscope (*AFM*) (Fig. **8b**) are similar. Practically, they have been developed simultaneously. A resolution of their measurement channel of a probe navigation is also comparable and may reach the hundredth fraction of a nanometer. Therefore, they are often joined in one complex or even in one head allowing the wider possibilities of probing, which makes them the most popular modern nanotechnologies. A combination of the *AFM* and the nanoindentation method allows a 2D surface studies under a normal and a lateral mode, and a 3D characterization of mechanical properties at a desired depth (from a unit to thousand of a nanometer). Fig. **9** shows a general view of a nanoindenter NANOINDENTER 11 (MTS Systems Inc., USA).

Figure 9: A photograph of a nanoindenter "NANOINDENTER 11" (the Bakhul Institute for Super Hard Materials, Kiev).

A typical experimental curve of a continuous indentation and a load dependence on an indentation depth are presented in Fig. **10**. An upper curve corresponds to a loading and reflects a material resistance to an indenter penetration; a lower one describes a deformation restoration after a load relief and characterizes the material elastic properties.

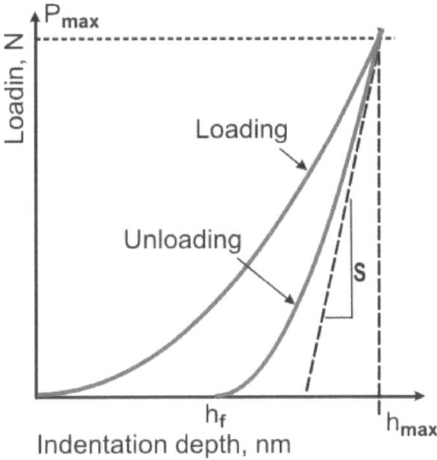

Figure 10: A dependence of an indentation depth on a load.

Under ordinary measurements of the material hardness, a main difficulty is to measure an indentation size under a low load. The main problem of the nanoindentation method is to treat an obtained diagram of a nanoindenter penetration. A device does not measure an indentation depth; it measures an indenter motion h_{max}, which is a sum of a contact depth h_c and an elastic deflection of a sample surface at a contact edge h_g [16]. To derive a nanohardness and an elastic modulus of a sample, with a diagram of an indenter loading, a contact depth h_c under maximum load P is to be known. The main difficulty is to find a surface elastic deflection at the contact edge h_g (Fig. **11**). The elastic deflection cannot be measured, it only can be determined from the following expression using a method of Oliver-Pharr [18]:

$$H_s = \varepsilon P_{max}/S, \tag{1}$$

where a coefficient $\varepsilon = 1$ for a flat die; $\varepsilon = 0.75$ for a paraboloid of a rotation and a sphere; $\varepsilon = 0.72$ for a sharp cone. A contact rigidity $S = dP/dh$ is derived from an indenter loading curve (Fig. **10**).

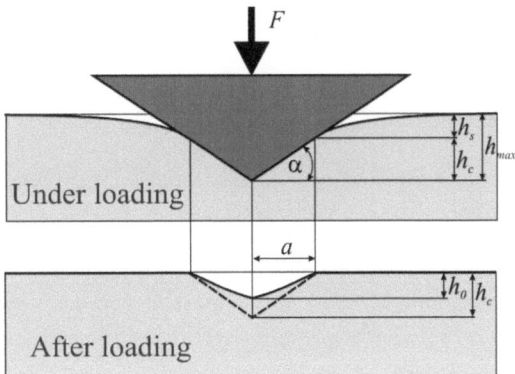

Figure 11: An indentation cross-section.

To find h_c using the Oliver-Pharr method for a Berkovich indenter, $\varepsilon = 0.75$ [19]. Knowing the contact depth, a projection area of indentation A can be found. For an ideal sharp Berkovich indenter:

$$A = 24.56 h_s^2 \tag{2}$$

Then, one can find the hardness over the indentation depth under a maximum loading using an expression:

$$H = P_{max}/A \qquad (3)$$

Using a spatial analysis and a method of finite elements, a ratio of work, which is spent for a plastic deformation (W_e) to a complete work W_c - $\dfrac{W_c - W_e}{W_c}$ can be calculated. This ratio characterizes a material plasticity λ, which can be calculated from the indentation curves (loading-unloading), when the indenter work is calculated as area under these corresponding curves. A dependence of the hardness ratio to the Young modulus (H/E^*) on the ratio of work spent for the plastic deformation to the work spent for the full one (a characteristic of a plasticity under indentation-λ) may be approximated as [17, 20, 21]:

$$\lambda = (W_c - W_e)/W_c = 1 - 5H^* \qquad (4)$$

Processing of the "P-h"-diagrams obtained from the nanoindentation allows [15]:

- Determination of a resistance to an elastic-plastic local deformation in a nanocontact area;

- Determination of $H = P/S$ hardness of the elastic-plastic contact (here P is a penetration strength, S is an indentation area related to its depth h through a geometry of an indenter vertex);

- Measurement of energy absorbed during a contact interaction;

- Determination of elastic-plastic characteristics of a material failing to be deformed plastically (a ceramics, a mineral and a metallic glass, a carbide, a nitride, a metallic boride, *etc.*);

- Determination of mobility characteristics for an isolated dislocation and aggregates in a crystalline material;

- Determination of a destruction viscosity coefficient K_{1c} from a crack size around an indentation area and a value of an indentation force;

- Modeling of a wear and fatigue process in a near surface layer by a many-time loading of the same region and nanocracking;

- Evaluation of material porosity;

- Study of a structure of a many-phase material;

- Study of a phase transition induced by high stress under an indenter;

- Determination of an elastic modulus, sound velocity and anisotropy of mechanical properties;

- Determination of thickness, an adhesion degree, and mechanical properties of a thin layer and a coating;

- Study of time-dependent characteristics of a material and velocity-sensitive coefficient of mechanical properties both during a stage of loading and during the stage of a tough-elastic restoration of an indentation after a relief;

- Measurement of a value and a distribution of an internal stress.

Being convenient and flexible, the nanoindentation method allows a study of mechanical properties of a solid in a thin near surface layer.

QUESTIONS FOR CONTROL

1. What is a high-resolution electron microscopy?

2. What methods are applied to study a microstructure of a solid surface?

3. What is a difference between a STM and an AFM method?

4. What method is applied to determine a surface chemical composition?

5. What characteristics of a nanomaterial can be studied by a nanoindentation method?

6. Remember a plot of a load dependence on a penetration depth. What does it characterize?

7. Enumerate mechanical characteristics, which can be determined using a nanoindentation method.

REFERENCES

[1] Protsenko IYu, Chornous AM, Protsenko SI. Equipment and Methods for Researches of Film Materials. Sumy: SumDU 2007.
[2] Feldman L, Mayer D. Fundamentals of Analysis of Thin Film Surfaces. Moscow: MIR 1989.
[3] Gladkikh NT, Dukarov SV, Kryshtal AP, et al. Surface Phenomenon and Phase Transformations in Deposited Films. Kharkov: KhNU 2004.
[4] Practical Scanning Electron Microscopy. Translation Ed. By Petrov V.I. Moscow: MIR 1978.
[5] Pines BJa. Lectures on Structure Analysis. Kharkov: KhGU 1967.
[6] Naumovets AG. Interaction of Fast Particles with Surfaces of Solids. Moscow: MIFI 1979.
[7] Anischik VM, Ponariadov VV, Uglov VV. Diffraction Analusis. Minsk: BGU 2002.
[8] Gusev AI. Nanomaterials, Nanostructures, Nanotechnologies. Moscow: Fizmatlit 2005.
[9] Oura K, Livshits VG, Saranin AA, et al. Introduction into Physics of Surface. Moscow: Nauka 2006.
[10] Azarenkov NA, Beresnev VM, Pogrebnjak AD. Structure and Properties of Protecvite Coatings and Modified Layers of Materials. Kharkov: KhNU 2007; 565.
[11] Woodraf D. Modern Methods for Surface Investigation. Translation. Moscow: MIR 1989.
[12] Kovaliov AI, Scherbedinskii GV. Modern Methods for Investigation of Surfaces of Metals and Alloys. Moscow: Metallurgiia 1989.
[13] Uglov VV, Cherenda NN, Anischik VM. Methods of Analysis of Element Composition of Surface Layers. Minsk: BGU 2007.
[14] Malygin GA. Plate-Type and Strength of Micro-and Nanocrystalline Materials. Phys Solid State 2007; 49: 951-982.
[15] Golovin YuI. Nanoindentation and Mechanical Properties of Solids in Submicrovolumes, Thin Near Surface Layers and Films. Phys Solid State 2008; 50: 2113-2140.
[16] Dub SN, Novikov NV. Nanohardness Tests of Solids. Journal Sverkhtverdye Materialy 2004; 6: 16-31.
[17] Firsov SA, Rogul TG. Theoretical (Limiting) Hardness. Reports of NAS of Ukraine 2007; 1: 110-114.
[18] Oliver WC, Pharr GM. An improved technique for determining hardness and elastic modulus using load and displacement sensing indentation experiments. J Mater Res 1992; 7: 1564-1583.
[19] Oliver WC, Pharr GM. Measurement of hardness and elastic modulus by instrumented indentation: Advances in understanding and refinements to methodology. J Mater Res 2004; 19: 3-21.
[20] Cheng Yang-Tse, Cheng Che-Msn Relationships between hardness, elastic modulus, and the work of indentation. Appl Phys Let 1998; 78: 614-619.
[21] Azarenkov NA, Kirichenko VG. Nucelar-Physics Methods in Radiation Material Science. Kharkov: KhNU 2008.
[22] Sung J, Lin J. Diamond Nanotechnology. Syntheses and Applications. Pan Stanford Publishing Pte. Ltd. 2010.

[23] Greim J, Schwetz KA. Boron Carbide, Boron Nitride, and Metal Borides, in Ullmann's Encyclopedia of Industrial Chemistry. Wiley-VCH: Weinheim 2005.

[24] Dubrovinskaia N, Dubrovinsky L, Solozhenko VL. Comment on "Synthesis of Ultra-Incompressible Superhard Rhenium Diboride at Ambient Pressure. Science 2007; 318: 1550c.

[25] Levine JB, Tolbert SH. Kaner RB. Advancements in the Search for Superhard Ultra-Incompressible Metal Borides. Advanced Functional Materials 2009; 19: 3519.

[26] Solozhenko VL. Ultimate Metastable Solubility of Boron in Diamond: Synthesis of Superhard Diamondlike BC_5. Phys Rev Lett 2009; 102: 015506.

[27] Levine JB, Nguyen SL, Rasool HI, Wright JA, Brown SE. Preparation and Properties of Metallic, Superhard Rhenium Diboride Crystals. J Am Chem Soc 2008; 130 (50): 16953.

[28] Levine, JB, Betts JB, Garrett JD, Guo SQ, Eng JT. Full elastic tensor of a crystal of the superhard compound ReB_2. Acta Materialia 2010; 58: 1530.

[29] Azarenkov NA, Beresnev VM, Pogrebnjak AD, *et al.* Fundamentals of Fabricated Nanostructured Coatings, Nanomaterials and Their Properties, Publ. House URSS, Moscow 2012; 352.

[30] Pogrebnyak AD, Sobol' OV, Beresnev VM, *et al.* Features of the structural state and mechanical properties of ZrN and Zr(Ti)-Si-N coatings obtained by ion-plasma deposition technique Tech Phys Lett 2009; 35 (10): 925-928.

[31] Pogrebnjak AD, Shpak AP, Beresnev VM, *et al.* "Structure and Properties of Nano-and Microcomposite Coating Based on Ti-Si-N/WC-Co-Cr" Acta Physica Polonica A 2011; 120 No. 1: 100-104.

CHAPTER 8

Structure and Properties of Nanostructured Films and Coatings

Abstract: Original results of studies of structure and properties of a nanostructured (nanocomposite) film and coating are presented. A classification of nanocrystalline films (3 groups) is presented. Effect of an ion bombardment in the process of deposition is considered. A nanocomposite coating (including multi-layered one), hard, and superhard one is considered. Their definition is presented and their thermal stability is considered. A whole class of super-hard composite materials featuring $H \geq 80GPa$ is described in detail. The super-hard nanocomposite is composed of a transition metal nanocrystal TiM (where M is Ti, Zr, Mo, Ta, *etc.*), which is surrounded by one mono-layer of an amorphous (or quasi-amorphous) system $\alpha\text{-}Si_3N_4$, BN.

Keywords: Amorphous, nanocompisite coatings, superhard, microstructure, mechanical properties.

8.1. INTRODUCTION

Ways, which are applied to form a condensed nanostructured material, are numerous. However, they are based on a mechanism of an intensive energy dissipation realized in three stages. During the first stage, a nucleus is formed, but due to certain thermo-dynamical conditions, it does not transit to a full-range crystallization. During the second stage, an amorphous cluster is formed around the nanocrystalline nucleus. During the third stage, they are united into an intercrystalline phase and form a dissipation nanostructure. A predicted modeling of such structure-phase condenstate state is difficult due to a presence of a thermodynamically metastable state. However, it opens new possibilities in the formation of a material with unique functional properties.

An application of new methods and approaches for a description of a non-linear process characterizing a dissipative state of the nanosystem seems to be an important step in understanding the features of the nanostructured state and in a control of a structure and properties "*in situ*" at the first stage of nucleation. In a linear system, a joined action of various factors is simple a superposition of results of every individual action. In a non-linear system, even an insignificant external action can lead to a great effect. Under a strongly non-equilibrium condensation from a high-energy ion-plasma flow, such action leads to the formation of a material with a unique structured state and properties [1-10].

An application of various ion-plasma (ion-vacuum-arc, magnetron) deposition methods seems to be the most promising for the fabrication of a nanostructured coating [11-46].

8.2. FORMATION OF NANOCRYSTALLINE FILMS

A nanocrystalline film is characterized by a wide, low-intensity reflection of an X-ray. Such films are formed in a transition region, when a crystal structure is strongly changed. There are three groups of the transition state [9, 10, 47-51]:

1. From a crystalline phase to an amorphous one;

2. Between two crystalline phases of different materials;

3. Between two preffered crystallographic orientations of grains of the same material (Fig. **1**) [33].

8.2.1. Role of Energy in Formation of Nano Structure Films

As it is known, a relation between solid properties and structures is of a fundamental importance not only for a material science but also for a physics of a thin film. However, process parameters of the film formation and its chemical composition determine what structure would be formed as a result. A problem is

Alexander D. Pogrebnjak and Vyacheslav M. Beresnev
All rights reserved-© 2012 Bentham Science Publishers

that every deposition process combines many deposition parameters, which are interrelated. Under magnetron sputtering, these parameters are a magnetron discharge current I_d, a voltage U_d, a substrate shift U_s, an ion current density of a substrate i_s, a substrate temperature, da istance from the substrate to a target d_{s-t}, a velocity of a sputtered gas flow φ, a partial reaction gas pressure p_{RG}, a full pressure of the sputtered gas $p_T = p_{RG} + p_{AR}$, a pumping-out rate of a pumping system, a basic pressure in a deposition chamber p_0, a position of entrance holes for the sputtered gas, a mutual orientation of a magnetron target and the substrate surface (a perpendicular or an inclined deposition), a stable, rotating or linearly moving substrate, a plasma enhance by an additional high-frequency, a superhigh-frequency or a full cathode discharge, an improvement of a plasma confinement by an external magnetic field usually formed by the electromagnetic Gelmgolz coils, and a deposition chamber geometry.

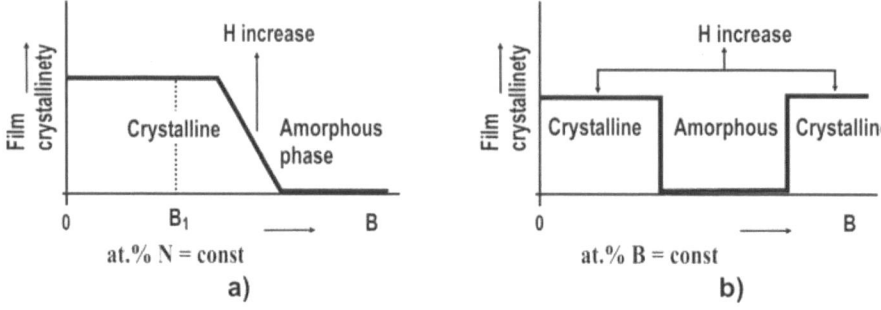

Figure 1: A schematic image of three transition regions: *a*-a transition from a crystalline to an amorphous phase; *b*-a transition between two crystalline phases or two dominating crystallographic grain orientations.

The most important deposition parameters for every sputtering device are I_d, U_s, i_s, T_s, d_{s-t}, a_D, p_{RG}, and p_T. Every combination of these parameters, however, is able to form only one discrete structure. Therefore, changing one parameter in this combination does not mean a continuous change of deposited film structure. This is the main reason why a film of desired structure and properties cannot be formed. A key helping to solve this problem seems to be a controlled energy E, which is delivered to a growing film. This energy may be delivered in three ways:

1. By a substrate heating T_s;

2. By an ion bombardment E_{bi} (an ion bombardment under a controlled ion energy E_i, an ion flow φ_I and a deposition rate a_D), and a fast neutron bopmbardment E_{fn} (a deposition using an atom with a controlled p_T, energy E_{ffp}, and v_{ffp} of film-forming particles);

3. By a chemical reaction ΔH_r (an exothermal reaction when $\Delta H_r < 0$, a heat is released and a full energy increases, and, on the contrary, by an endothermic reaction, when $\Delta H_f > 0$, the heat is consumed and the full energy decreases); here ΔH_f is an energy needed to form a desired composition.

All three components of the full energy influence a growing film simultaneously, but an effect of an individual component may strongly differ. For example, when a pure metal is deposited, the energy contribution from the chemical reaction is zero. On the contrary, when a film is deposited using the ion deposition process, when the growing film is bombarded by the low-energy ion, E_{bi}-the delivered energy plays a crucial role in its growing. Therefore, the ion bombardment is very often applied to govern properties of the deposited film.

The energy E_{bi} delivered to the growing film by the bombarding ion plays a very important role in its structure, physical, and functional properties. An energy of a collisionless discharge may be determined from three easily measured values, *i.e.* a substrate shift U_s, an ion current density at a substrate i_s, and a film deposition rate a_D, according to the following formula:

$$E_{bi} = E_i v_i / v_m = e(U_p - U_s) v_i / v_m \propto eU_s i_s / a_D \text{ under } T_s = \text{const,} \quad (1)$$

where E_i is an ion energy, v_i and v_m is an ion flow bombarding a growing film and covering atoms, respectively, and U_p is a plasma potential. Typical U_s value varies from those of a floating potential U_{fl} to approximately 200 V. To perform an efficient control for the film microstructure, one needs a value $i_s \geq 1$ mA/cm^2.

Every material can be characterized by a definite critical energy value E_c. A film formed under $E_{bi} < E_c$ is porous, soft, has a mat coating, and is under tension. On the contrary, a film formed under $E_{bi} > E_c$ is compact, dense, has a smooth surface, and demonstrates a high reflection ability, and a work for compression. A films formed under $E_{bi} = E_c$ demonstrates a zero macrostress σ. However, equal E_{bi} values do not mean a similar film microstructure. According to an equation (1), equal E_{bi} values may be reached under various combinations of an ion energies E_i and a ratio v_i/v_m, i.e. under conditions when various physical processes may dominate. This means that E_i and v_i/v_m parameters as the physical terms are not equivalent.

In the case of a collision discharge [11-13, 42-47], E_{bi} energy delivered into a unit of a deposited film volume may be expressed in the following form:

$$E_{bi} (J/cm^3) = U_s (i_s/a_D) Ni_{max} \text{ under } T_s = \text{constant,} \quad (2)$$

where $Ni_{max} = exp(-L/\lambda_i)$ is an amount of ions reaching a substrate with a maximum energy eU_s, e- is an electron charge, L is a film thickness, λ_i is an average free path of an ion for a collision resulting in a loss of an ion energy in a layer. An average free ion path may be calculated using a Dalton's law as $\lambda_i \approx 0.4/p$. A thickness L of a wire under high-voltage $(U_s >> U_{fl})$ may be calculated using a Charles-Langmuir law for the wire under d_c, where U_{fl} is the floating potential. The film thickness L may be expressed in the following form:

a) A collisionless layer under dc occurring near a substrate $(L/\lambda_i < 1)$

$$L = (0.44\varepsilon_0)^{1/2} (2e/m_i)^{1/4} U_s^{3/4} i_s^{1/2} \quad (3)$$

where ε_0 is a free volume dielectric penetrability and m_i is an ion mass.

b) A collision layer under dc occurring near a substrate $(L/\lambda_i > 1)$

$$L = (0.81\varepsilon_0)^{2/5} (2e/m_i)^{1/5} \lambda_i^{1/5} U_s^{3/5} i_s^{2/5} \quad (4)$$

This simple analysis indicates that an energy delivered to a growing film by the ion bombardment strongly depends on conditions accompanying the film sputtering. The energy E_T, which is delivered by the ions to a negative electrode (a substrate) and an amount of ions Ni_{max} reaching the negative electrode with a maximum energy eU_s as a function of L/λ_i are shown in Fig. 2.

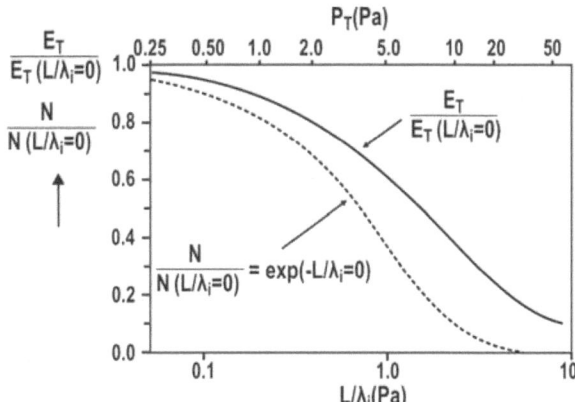

Figure 2: An amount of ions Ni_{max} reaching a substrate with a maximum energy eU_s and an energy E_T, which is transferred to a growing film as a function of L/λ_i.

The energy E_T decreased with increasing L/λ_i, i.e. with an increasing pressure of a sputtered gas p_T because $\lambda_i = 0.4/p_T$, and this means that $L/\lambda_i \sim p_T$ [12]. A decrease of the bombarding ion energy with increasing p_T affects essentially a mechanism of growing and a structure of the deposited film. This fact fairly well explains why properties of the film, which is deposited under the same U_s and i_s values, differ for the deposition under different p_T values.

When a negative potential is applied to the substrate, working gas ions can bombard the growing surface and perform an additional energical stimulation of the process.

Many researches used the equation (1) to characterize an effect of the low-energy ion bombardment on a film microstructure and properties. In spite of relatively wide applications of the ion deposition process, there are certain data concerning a correlation between the energy E_{bi} and properties of a reactively deposited film. Regardless of the fact that the reactive film sputtering is accompanied by a target etching (a cathode), which sharply decreases the film deposition rate, not many people can understand that a change of a_D induced by a change of a partial pressure of a reaction gas p_{RG} ($RG = N_2$, O_2, CH_4, etc.) under constant deposition conditions can crucially change E_{bi} energy delivered to a film during its growing. For example, for a nitride a_D (Me) $\approx 4a_D$ (M_eN_x) and for an oxide a_D (M_e) $\approx$ (10 to 15) a_D (M_eO_x), where M_e-is a metal, and M_eN_x, and M_eO_x is a metal nitride and an oxide, respectively. Therefore, we should take into account that a change of properties of a reactively sputtered film is induced by a simultaneous action of two parameters: (1) an element and a chemical film composition, especially by an amount of reaction gas atoms implanted into a film and (2) E_{bi} energy, i.e. those parameters, which depend on the reaction gas partial pressure p_{RG}. Under reactive sputtering, an effect of the increased E_{bi} due to a_D decrease when pN_2 increases may be very strong (Fig. **3**).

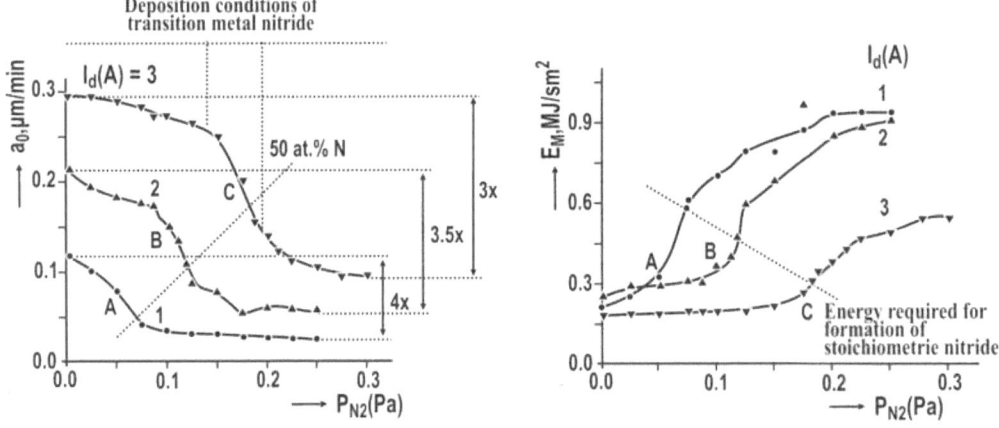

Figure 3: A deposition rate a_D for Ti (Fe) N_x film as a function of a nitrogen partial pressure pN_2.

It is assumed that only a decrease of a_D when p_{RG} increases is responsible for a sharp change of a crystallographic orientation of a single-phase film, which is based on a solid solution, for example Ti (Fe) N_x film.

E_{bi} energy strongly affects not only the film structure, its element and chemical composition (for example, due to a desorption of reaction gas atoms from a sputtered film surface), but also a macrostress σ induced in the film by the ion bombardment [6, 15, 17, 44-50].

It is known, that the ion bombardment is a highly non-equilibrium process, in the course of which an ion transfers its kinetic energy to a growing film and heats it at an atomic level. This heating essentially differs from a usual one. A kinetic energy of the bombarding ion is transferred to a very small region of an atomic range and is accompanied by an extremely prompt (approximately 10^{14} K/s) quenching. In addition, one should note that the energy delivered to a growing film by an ordinary heating (T_s/T_m) and a particle bombardment (E_{bi}) are not physically equivalent. Here, T_s is a substrate temperature and T_m is a melting point of the film material. A dense film with extraordinary properties, which corresponds to T-zone in a structure zone model of Thornton, can be formed, if the sputtering is performed under low pressure of approximately 0.1 Pa and lower (Fig. **4**).

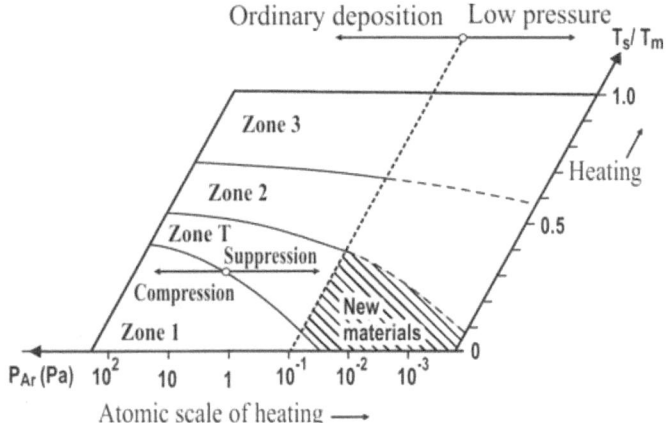

Figure 4: A Tornton model for a sputtering region under low pressure as a way to form a new advanced material under low deposition temperature T_s.

The low-pressure sputtering shifts the transition zone T to a region of low-value T_s/T_m ratio. This allows the formation of a dense film corresponding to T zone under low deposition temperature T_s.

The energy E_{bi} transferred by the bombarding ion to the growing film essentially affects the film structure, physical, and functional properties. All of the above-mentioned concerns a magnetron sputtering, when an energy of a deposited particle is regulated by the working gas pressure in the vacuum chamber, by a distance to a substrate, and by a negative potential applied to the substrate.

In the case of a vacuum-arc method and the ion deposition, an energy of the deposited ion can be controlled by a change of an accelerating potential value, which is applied to the substrate. In this way, a flow density in the course of coating formation can be regulated. A change of the deposited particle energy affects a material structure and a substructure of an ion-plasma deposited film.

8.3. FEATURES OF FORMATION OF NANOCRYSTALLINE COATINGS

A desired size and crystallographic orientation of a growing film can be regulated by:

- An application of an ion bombardment in the process of coating deposition;

- An implantation of an additional element, which is able to limit the grain size, into a basic material ;

- A multilayer film deposition when thickness of every layer is of a nanometer range;

- A formation of a nanocomponent coating.

8.3.1. Effect of Ion Bombardment on Coating Formation

Earlier, the works of Musil *et al.* reported that one of the ways to change a microstructure, physical, and mechanical properties of a coating was a bombardment of a growing surface by energetic ions, which was realized in the process of deposition. In this case, the ion bombardment decreases a crystallite size, makes a grain interface denser, induces defects (Frenkel pairs and other point defects), and a compressing stress.

For example, when TiN coating is deposited using a vacuum-arc deposition, a structure element size of the coating can be decreased by an application of a negative voltage pulse of 1 to 2 kV, 1 to 7 Hz frequency, and a constant voltage of 0 to 500 V to a substrate in the course of deposition. Fig. **5** illustrates a fractography image for TiN coating.

It is seen, that the formed coating has a columnar structure, which is characteristic for the ion-plasma deposited coating. An evaluation of the crystallite size according to a width of an X-ray line gives the average values of 15 to 30 nm. At the same time, the average size of TiN crystallite without the implantation under the constant bias voltage is 100 to 200 nm. In such a way, the ion bombardment, which is realized in the course of coating formation, decreases the grain size and changes the structure and properties of a resulting material.

a)
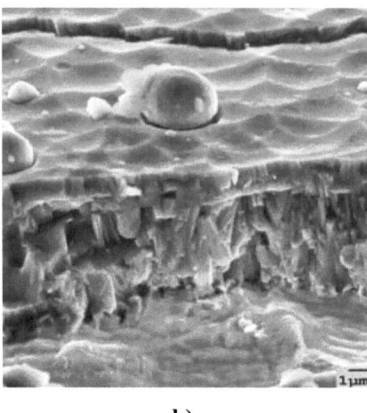
b)

Figure 5: Fractography images of fractures found in TiN coating, which is deposited under 0.66 Pa nitrogen pressure: *a*-under a negative constant bias of 230 V and 2 kV impulse; *b*-under a negative constant bias of 230 V [1].

8.3.2. Process of Mixing

A mixing is an addition of one or several elements to a basic material composed of one element. An introduction of the doping element prevents growing of grains of a basic coating phase. Fundamental parameters applied to control the film structure are a substrate temperature T_s, an energy E_{bi} transferred by a bombarding ion and a fast neutral-charge ion to a growing film, and an amount and a type of a doping element. However, other factors also play an essential role in the formation of a nanocrystalline film:

- An ability or an inability of film elements for a mutual mixing;

- An ability of elements to form a solid solution or an intermetallic compound;

- An enthalpy in the alloy formation ΔHf (negative or positive).

What a film structure will be formed as a result strongly depends on these factors and their combination. So for example, a microstructure of a single-phase film qualitatively is fairly well described by a model proposed by Movchan and Demchishin, and Thornton. However, all these models can violently change if a film is doped by an impurity element. The impurity or a doping element stop the grain growing and stimulate a re-nucleation. This phenomenon results in the formation of a globular structure, which stretches both to a low and high T_s/T_m value, depending on the impurity or the doping element content in a film. Here, T_s is a deposition temperature and T_m is a melting temperature of the coating material (Fig. **6**). When the content of impurity or doped element is average or high, a columnar structure, which is typical for *I* zone of a one-phase film, totally disappears. This fact is described by a model developed by Barn and Adamic. A two-phase nanocomposite coating demonstrates a microstructure, which is identical to that predicted by the Barn-Adamic model (Fig. **6a**). In the figure, Zr-Cu-N nanocomposite film structure with a low (1.2 at.%) and high (20 at.%) Cu content is compared to the model. We should mention that even 7 at.% of Cu is enough to form a dense small-grained Zr-Cu-N film without the columnar microstructure. A similar transformation of the columnar microstructure into a dense small-grained one is observed in TiN film after a deposition of Si atoms, *i.e.* for Ti-Si-N films.

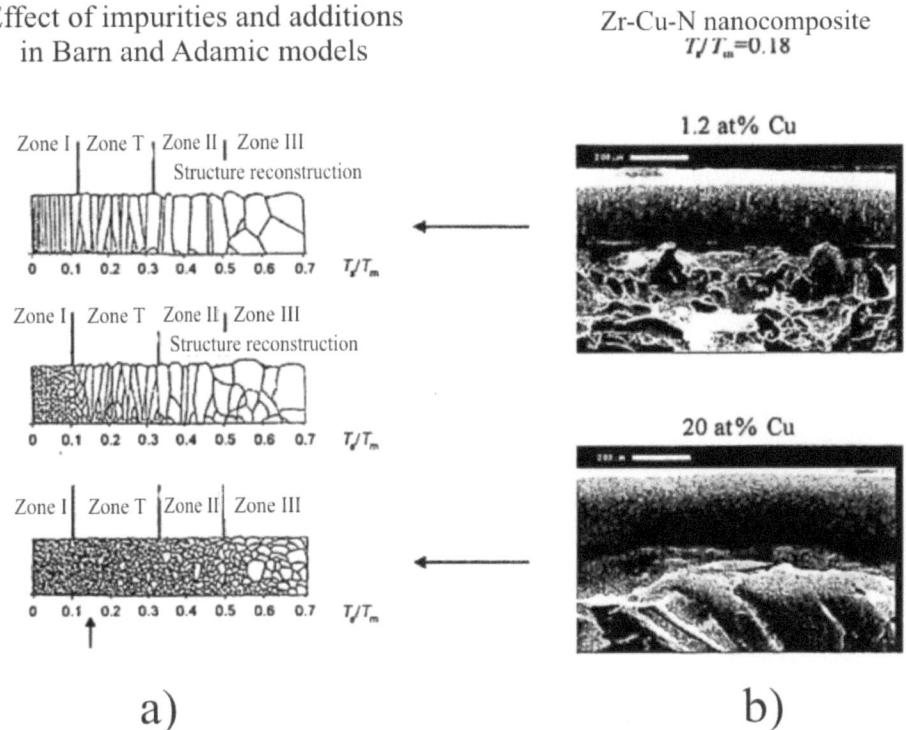

Figure 6: A comparison of Barn-Adamic model with experimental data: *a*-a model for a double-phase film; *b*-a cross-section of Zr-Cu-N film with a low (1.2 at.%) and a high (20 at.%) copper content.

Ti-Si-N film with a low (< 5 at.%) Si content has columnar microstructure, a film with 11 at.% of Si and 14.9 at.% of Si are not columnar but composed of a nanocrystalline equal-axed TiN grain. In addition, a transition from the columnar structure to the nanocrystalline equal-axed microstructure corresponds to a transition from a crystalline to an X-ray-amorphous phase. In a binary Ti-Si alloy, already Ti-Si film with approximately 12 at.% of Si indicates several diffusion diffraction peaks. It means that this film contains a mixture of grains with different crystallographic orientations. The binary Ti-Si film with approximately 12 at.% of Si or higher is amorphous. Zr-Si-N film demonstrates the same behavior. It turns out that the columnar microstructure disappears, if an amount of Si added to a basic ZrN material exceeds approximately 5 at.%.

This indicates that the transformation of the columnar structure takes place also in the case, when a one-phase material is transformed into a two-phase one, for example by a change of its chemical composition. This phenomenon is demonstrated for WC-Ti$_{1-x}$Al$_x$N film. The film with x = 0.3 demonstrats the pronounced columnar microstructure. The film with $x = 0.57$ is very uniform and does not demonstrate the columnar microstructure. A stoichiometry x = 0.57 corresponds well to Al concentration range between 50 and 60 at.%, in which the two-phase film composed of TiN and AlN grain mixture can be formed. This change in the film stoichiometry corresponds fairly well to the transition from the crystalline to the amorphous phase of a binary Ti-Al alloy. Namely, it was found that Ti-Al alloyed elements containing 35 to 59 at.% of Al were X-ray-amorphous. This indicates that the film with a dense small-grain structure may be formed not only by a dopinf of an impurity and/or by an addition, but also by a selection of such deposition conditions, which would allow the formation of a film containing a mixture of nanocrystalline grains of different materials, with various crystallographic orientations, and/or various lattice structures and grains with a strong dominating crystallographic orientation.

A key role in the formation of the nanostructured film is played by an energy, which is transferred to a film during growing. In comparison with the ion deposition, this process allows the formation of a film containing a mixture of various nanocrystalline grains and nanoamorphous films with a nanocrystalline structure [18-26, 39-47].

8.3.3. Nanostructured Multilayered Coatings

An efficient way to control the crystallite size during growing is the formation of a multilayered nanostructure. The multilayer composition can be formed by a periodical deposition of an individual thin layer of various refractory compounds of a desired thickness. In this case, a fraction of phase interfaces in this nanomaterial increases with respect to a summary interface volume. This strongly affects properties of the multilayered coating. The grain interface is an obstacle for a propagation of dislocations and cracks. Namely, this essentially increases the coating hardness. Fig. 7 illustrates nanostructured $TiN_x/Cr_x/AlN$ coating and a change of the hardness depending on an alternation of layers.

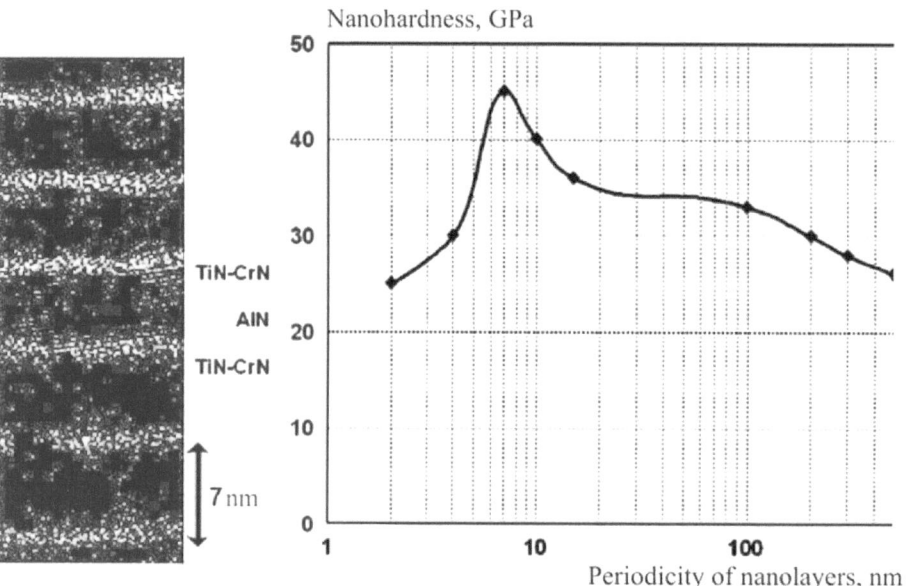

Figure 7: A microstructure and a hardness of a multilayered coating.

It is seen, that a high hardness value can be reached at a definite alternation of layers and a definite thickness of a nanolayer. When the thickness is low (6 to 7nm) the hardness decreases due to a smooth layer interface. The formation and studies of such coating is of a high scientific and practical interest.

8.3.4. Nanocomposite Coatings

The nanocomposite coating represents a new generation of materials. The nanocomposite coating, which drew a high interest in the middle of 90th of XX century, strongly differs from an ordinary material. The ordinary material is a directed-granulation material with a bigger than 100 nm grain size (d). An improvement of its properties is based mainly on doping of a basic material. A new crystalline superalloy and superceramics is formed in this way. Due to a relatively big grain d, a deformation process in the ordinary usual material complies an influence of dislocations, *i.e.* a process of their formation and motion. Namely this process determins the basic properties of a volume material and coating, for example, a hardness H, a Young modulus E, a plastic deformation, an elastic recovery, a viscosity (a strength), a crack resistance, a heat stability, an oxidation resistance, *etc.* The dislocation activity is the main reason why the properties of the ordinary material, which is composed of the big grain (> 100nm) may be only improved by an element doping, but cannot be radically changed like a nanocomposite material with a small grain (< 100 nm). A material containing a mixture of two different kinds of the small grain (< 100 nm) can be determined a *nanocomposite material*. A dislocation fails to be generated in a grain of $d \approx 10$ nm size.

This means that when the grain size d decreases, the dislocation activity is gradually changed by a new deformation process, in particular an enhancement of the grain interface, a sliding of the grain interface, and an electric binding between an atoms of the adjacent grains and/or atoms of the adjacent regions. In addition, a ratio S/V of a surface S and a volume V and a ratio N_b/N_g, which is an atomic amount in the grain interface region and inside the grain itself, increase strongly with a decreasing d. Properties of the nanocomposite material are determined by the grain size, form and a volume of the grain interface. This is the main reason why the nanocomposite coating demonstrates the improved properties and very often, some unexpected unique physical and functional properties.

Due to (1) the very small grain size (< 10nm) and (2) a significant role of interface regions surrounding an individual grain, the nanocomposite material behaves in a different way, in comparison with the ordinary material, the grain of which is bigger than 100nm, and demonstrates absolutely new properties. This special class of materials is characterized by a heterogeneous structure formed by practically non-interacting phases having < 100 nm average linear size of the structure elements. They are composed at least of two phases the nanocrystalline and the amorphous structure. Today, the most pronounced progress in studies is observed for the systems, which demonstrates a total or a practically total impossibility to mix the components, because the solid nanocrystallite is fully surrounded by a material of a different phase, which is in an amorphous state.

Veprek *et al.* proposed a theoretical concept for the formation of the solid nanocrystalline nanocomposite coating. According to this concept, the coating contains a dislocation-free nanocrystallite (a solid phase) of 3 to 10 nm size, separated by an amorphous intermediate layer composed of 1 to 3 nm grains.

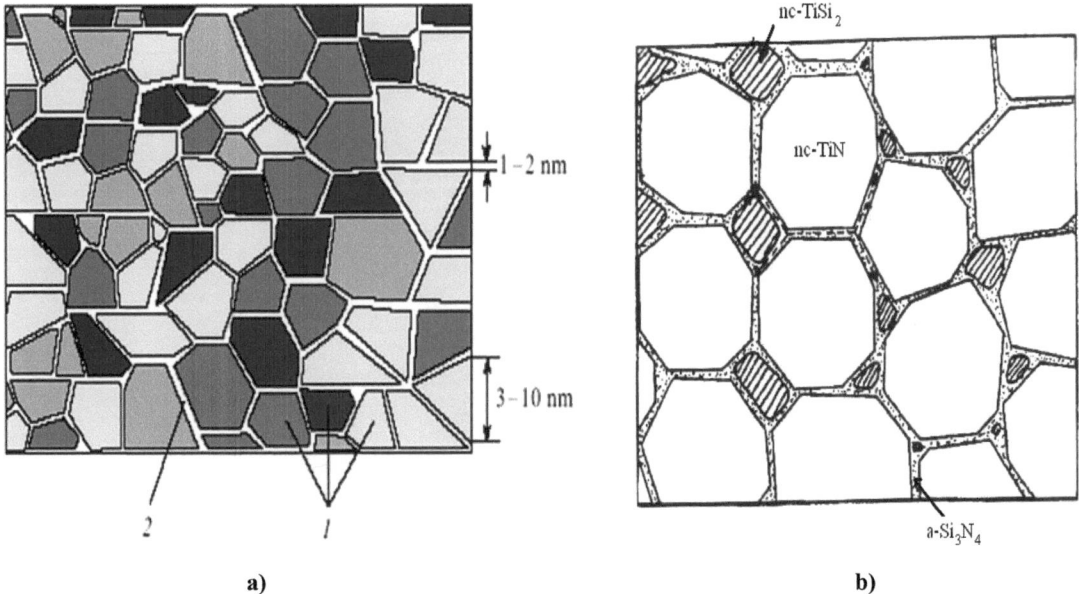

Figure 8: A schematic image of a nanocomposite structure: *a*-for an ideal nanostructured coating; *b*-a structure of *nc*-TiN/α-Si$_3$N$_4$/*nc*-TiSi$_2$ nanocomposite [6, 8, 9].

An ideal model of the superhard nanocomposite coating is demonstrated in Fig. **8a**. Fig. **8b** shows a schematic image of Ti-Si-N nanocomposite.

Today, several nitride systems are known. A system TiN-Si$_3$N$_4$ is studied most thoroughly.

The composite material TiN-Si$_3$N$_4$ formed by a magnetron sputtering under an average deposition rate 0.2 to 0.3 nm/s and an atmosphere working pressure $P_{Ar+N} \approx 0.3$ to 0.4 Pa demonstrates a stable hardness at about 900K temperature for a long time. Its hardness is close to a hardness of a boron nitride in a bulky sample state.

8.4. NANOCRYSTALLINE COATINGS OF HIGH HARDNESS

The basic mechanisms, which are responsible for the increased film hardness H are the following: a-a plastic deformation with a dislocation domination; b-a binding power between atoms; c-a nanostructure; d-a compressing macrostress σ generated in the film in the process of its formation. The value of material hardness H depends on the deformation processes working in a given interval of a sizerange d (Fig. **9**) [33-37, 48-50].

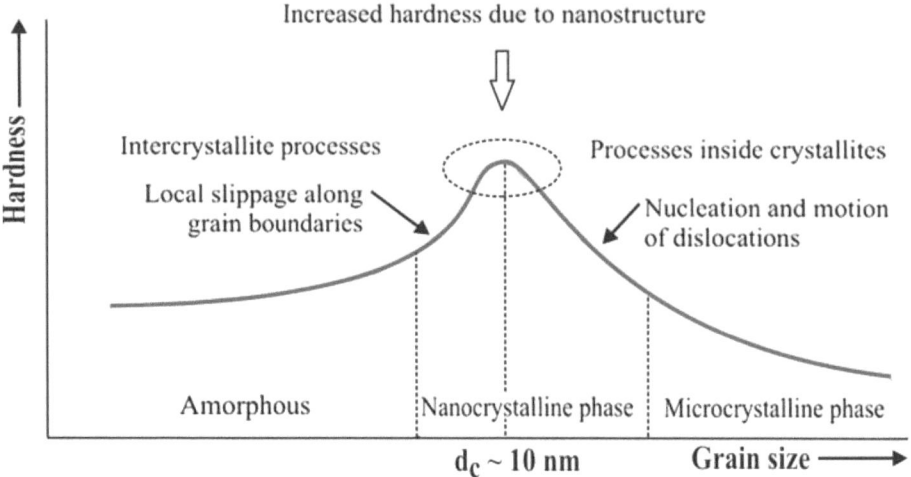

Figure 9: Dependence of material hardness H on grain size [9].

The maximum hardness H_{max} is found for $d_c \sim 10$ nm. A region near H_{max} corresponds to a continuous transition from a microscopic range of nucleation and a dislocation motion when $d > d_c$, which is described by the known Hall-Petch law ($H \sim d^{1/2}$) for an ordinary polycrystalline material, to a local range of an intercrystalline slipping through the grain interfaces and phases under $d < d_c$.

However, an idea to reach an additional increase of the hardness by transferring a refractory material into the nanostructured state is not so simple.

Among two basic types of hard condensed nanomaterials-a multilayered and a composite one, the latter demonstrates a higher increase of the hardness in comparison with the bulky sample state. Today, only two groups of nanocomposite coatings are known:

1. *nc*-MeN/soft phase and *nc*-MeN/solid phase;

2. A nanocrystalline or an amorphous phase containing two crystalline phases; with two crystallographic orientations of grains of the same material and a great difference in a microstructure of one of two phases.

The nanocomposite coating are subdivided according to three basic criteria: 1-the hardness; 2-the phase composition; 3-the size of a nanocomposite individual phase.

   i.   Classification according to hardness:

a)   a hard coating with $H \leq 40$ GPa;

b)   a superhard coating with $H = 40$ to 80 GPa;

c)   an ultrahard coating with $H \geq 80$ GPa.

ii. Classification according to a phase:

a) Two solid phases of nc-MeN/a solid phase, for example, a-Si_3N_4, BN, etc.;

b) One solid and one soft phase of nc-MeN/a soft phase, for example, Cu, Ag, Au, Ni, Y, etc.

Here nc means a nanocrystalline phase, Me = Ti, Zr, Ta, Mo, W, Cr, Al, and others are a nitride forming element. Both composites demonstrate the high hardness.

iii. Classification according to a phase size:

a) 2D-a two-dimensional coating; a superlattice coating;

b) 3D-a three-dimensional coating: a single-layer nanocomposite coating.

The superlattice coating is composed of two layers of a different composition, which are alternated many times. A summary thickness d of these two layers is called a superlattice period. The superlattice period varies from several nm to about 15nm. The three-dimensional single-layer nanocomposite coating up to several μm thickness is composed of nanograins (phase 1). The nanograins are either implanted into a matrix or coated by a thin layer, which is equivalent to a basic material (phase 2). A concept allowing the formation of a superhard nanocomposite material is based on this knowledge. Let us consider this concept, which is followed by Prof. Musil and co-authors. This concept is based on an unusual geometry of the nanostructure, $i.e.$ a grain size and a crystalline form. The nanostructure of the nanocrystallite with an increased H may be different. There are three types of a nanocomposite film microsctructure featuring the increased H: a) a columnar structure; b) a nanograin surrounded by a very thin phase of a basic material (~ 1 to 2 monolayers), and a mixture of nanograins with various crystallographic orientations (Fig. 10) [34-36].

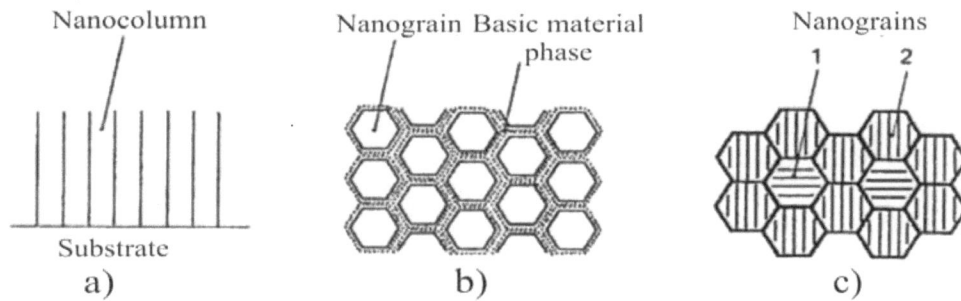

Figure 10: A schematic image of various nanocomposite structures with an increased hardness: a-a columnar structure; b-nanograins surrounded by a basic material phase; c-a mixture of nanograins.

According to the film nanostructure, the nanocomposite material with the increased H may is divided into three groups:

1. A nanocomposite material with a columnar structure, which is composed of grains united in a nanocolumn. In this case, an amount of a second phase (a basic material) is not enough to cover all grains, Fig. **10a**.

2. A nanocomposite material with a dense globular nanostructure containing nanograins, which are totally surrounded by a thin phase of a basic material, Fig. **10b**.

3. A nanocomposite material with a dense globular nanostructure composed of nanograins of various materials (a two-phase material) or nanograins with various crystallographic orientations and the same lattice structure (a one-phase material), Fig. **10c**.

The nanocomposite material formed in a crystalline interface (Fig. **10a** and **10b**) also demonstrates a columnar structure. The nanocomposite material containing nanograins totally surrounded by a basic material phase is formed in the process of a transition from a crystalline to an amorphous state (Fig. **10a**). The nanocomposite material containing a mixture of small nanograins of various materials or nanograins of various crystallographic orientations and/or the same lattice structures is formed in an inside of two crystalline phases or two dominating crystallographic grain orientations.

The above classification is confirmed experimentally. The increased H is closely connected to a size and a form of a standard block, which forms a nanocomposite material. Taking into account this fact, we can conclude that both the block geometry and the grain size are those physical parameters, which determine the new unique properties of the nanocomposite film. The increased hardness is found in a coating containing nanograins of the same material but with various crystallographic orientations and dirrefent lattice structures. This fact explains the increased hardness of the one-phase material. Today, a system with a full or practically full mixing disability of components and a system, in which a solid nanocrystallite is totally surrounded by a material of a different amorphous phase, can be successfully fabricated. Several nitride systems are now suroughly studied. $TiN-Si_3N_4$ system is the most popular one.

8.5. MECHANICAL PROPERTIES OF NANOCRYSTALLINE COATINGS

There are three parameters, which lead to the increased hardness of a solid nanocomposite material:

1. A macrostress σ, which is generated in a growing coating;

2. A nanocomposite nanostructure;

3. A short covalent atomic bond existing, for example, in Si-C-N and Si-C-B-N coating.

The increased hardness is a result of a simultaneous action of these two or even three parameters. This fact significantly complicates a correct determination of a real origination of the increased hardness and a selection of an optimal combination of deposition conditions, which could lead to the maximum hardness. Therefore, the easiest way is to study the coating hardness H as a function of only one parameter is the elimination of two from three existing coeficients. A very high compressing macrostress σ is able to detach a film from a substrate, if the film thickness h is higher than a critical value. Therefore, we should, at least, try to decrease σ. The macrostress may also be eliminated by an accurate control of the coating deposition parameters. Table **1** presents the hardness of nanofilms, which are formed using various deposition methods.

Table 1: Hardness of nanostructured coatings.

Composition	Method of Formation	Thickness, μm	Grain Size, nm	Hardness, H, GPa	Notes
TiN	Magnetron sputtering	1 to 2	5 to 30	35 to 40	
TiB_2	Magnetron sputtering	1 to 4	2 to 8	50 to 70	
Ti(B, N, C)	Magnetron sputtering	5 to 12	1 to 5	60 to 70	
$TiN-Si_3N_4$-$TiSi_2$	CVD	3.5	3.0	~ 100	
TiN/VN	Magnetron sputtering	2.5	2.5	54	
TiN/NbN	Arc deposition	2.0	10	78	
TiN/MoN	Arc deposition	Periodical, 9.6 nm		42 to 48	Deposition using pulsed HF
Ti-Si-N	Arc deposition	10	15 to 20	43.7	Cathode, sintered powder component Ti-15%Si

TiN	Arc deposition	3.5	15 to 20	40 to 68	Deposition with implantation
TiN/CrN	Arc deposition	Periodical, 8.7 nm		50	
TiN/CrN	Arc deposition	Periodical, 8.7 nm		60	Deposition with implantation

Analyzing the results presented in Table **1**, we can conclude that the hardness of a coating containing a hard material can be reached in two ways. (1) An energetic ion bombardment during a deposition process (TiN, TiN/CrN, TiN/MoN) induces a densification of a grain interface, decreases a crystallite size, and increases the hardness due to defect formation, as well as generates a high biaxial compressing stress. (2) A corresponding nanostructure prevents a growth, a multiplication, and a propagation of a deformation like a microcrack and a dislocation (leading to a change in a crystal lattice volume) and forms a nanocomposite coating of a definite composition. Fig. **11** shows a dependence of the hardness change for a nanocomposite nc-TiN/α-Si$_3$N$_4$ coating as a function of a silicon nitride content.

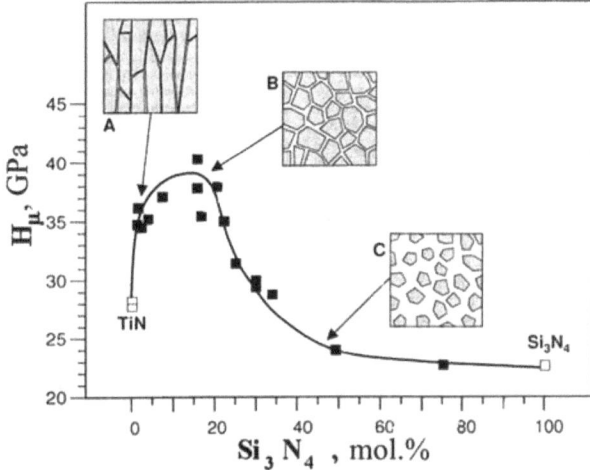

Figure 11: An effect of a silicon nitride content on a material hardness for nc-TiN/α-Si$_3$N$_4$ system [19, 37].

According to an electron microscopy studies performed for various contents corresponding to a different material hardness, a titanium nitride without a silicon contains individual TiN crystallites of several nanometer size, which are directed perpendicular to a growth plane and crystallites of several hundred nanometer size, which are directed parallel to the growth plane (an Inset A in Fig. **11**). An implantation of a small silicon amount during deposition essentially increases the hardness. In this case, even a partial coating of a titanium nitride grain by a silicon nitride prevents a further growing of TiN grain and stimulants a generation of a new TiN phase nucleus in the process of continuous deposition. When Si$_3$N$_4$ content is 15 to 20mol.%, an average TiN grain size does not exceed 7nm. This size is too small to activate a dislocation source, which results in a decrease of plasticity and an increase of the material hardness. Such material can minimize an action of mechanical loading only if a grain will slip along an interface (*i.e.* if an individual non-deformed TiN nanocrystallite will move as one relative to another). This process requires more energy than a deformation induced by a dislocation motion, which results in an increase of the material hardness. An evaluation of an average distance between a titanium nitride grain demonstrates that maximum high mechanical characteristics is reached if TiN nanocrystallite is detached by several silicon nitride single-layers (an Inset B in Fig. **11**). When an amount of the silicon nitride is high, an average distance between grains is sufficient for an initiation and development cracks in Si$_3$N$_4$ phase. In this case, the material hardness is close to the hardness of a bulky Si$_3$N$_4$ (an Inset C in Fig. **11**).

An evolution of the nanocomposite material hardness demonstrates that two important factors increase the hardness of such materials. A crystallite size cannot be lower than 10 nm in a deformation direction in order that to prevent a dislocation motion. An average distance between grains cannot exceed 0.5nm in order that to prevent the crack nucleation and the development.

A material state inside the grain interface, which, according to an electron microscopy data, is a non-equilibrium amorphous-like structure, is very important in this case. A stress long-range field of a non-equilibrium interface is characterized by a deformation tensor. Its components inside the grain are proportional to $r^{-1/2}$ (r is a distance to a grain interface). Therefore, the stress long-range field elastically distorts a crystal lattice. A value of these elastic distortions is maximum in the vicinity of the interface. In addition, a triple junction of ultrasmall grains may be considered as a declination. It is a paracrystal immersed in a medium with randomly packed atoms. Fig. 12 shows a calculated dependence of a volume fraction of grain interfaces and triple grain junctions on a grain size d_3.

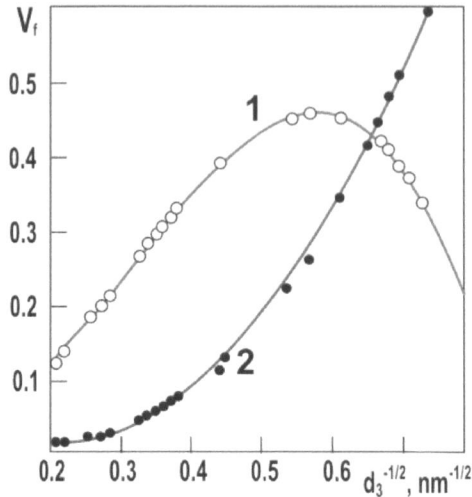

Figure 12: A calculated dependence of a volume fraction V_f of grain interfaces (1) and triple junctions (2) on a grain size (an interface thickness is taken to be equal to 1 nm in this calculation).

A comparison of the data obtained from studies of a relation of the mechanical characteristics to a change of the interface and triple junction density when the grain size decreases demonstrated that an extremum, which appears in a dependence of mechanical properties on the grain size under a characteristic size of a structure element 8 to 10nm, correlates well with the formation of a large volume fraction of the grain triple junctions. As a result, the material gains a shear instability and behaves as an amorphous one.

In such a way, due to a large fraction of phase interfaces, the nanocrystalline material is thermodynamically non-stable and tends to transform into a usual big-grain material with a small amount of phase interfaces. Therefore, a stabilization of the nanocrystalline grain structure plays a crucial role in a saving of the nanophysical and nanomechanical properties.

The main characteristic of a construction material is a Young modulus, a yield point, an ultimate strain, a fatigue limit, a wear resistance, and a failure toughness (a viscosity) (a critical coefficient of a strain intensity for sharp concentrator and a crack K_{si}). In contrast to the Young modulus, which in the first approximation does not depend on a material structure, all the rest characteristics are structure sensitive, *i.e.* may be controlled by a purposeful change of the real structure, in particular, a nomenclature and a structure defect concentration, a size of a grain, a cell, or another substructure unit.

Fig. 13 illustrates a dependences of H on E for a superhard coating formed by a magnetron sputtering. It is seen, that the hardness of a multicomponent nanostructured coating is higher than a single-layer one based on the nitride. However, in spite of a relative simplicity of a hardness study, we should always pay attention to an interval of an applied load, a film thickness, a surface topography, a residual stress, as well as the other factors, which can influence the hardness.

In most cases, on one hand, a decrease of the structure element size and a layer thickness to nanorange improve the mechanical properties of coatings.

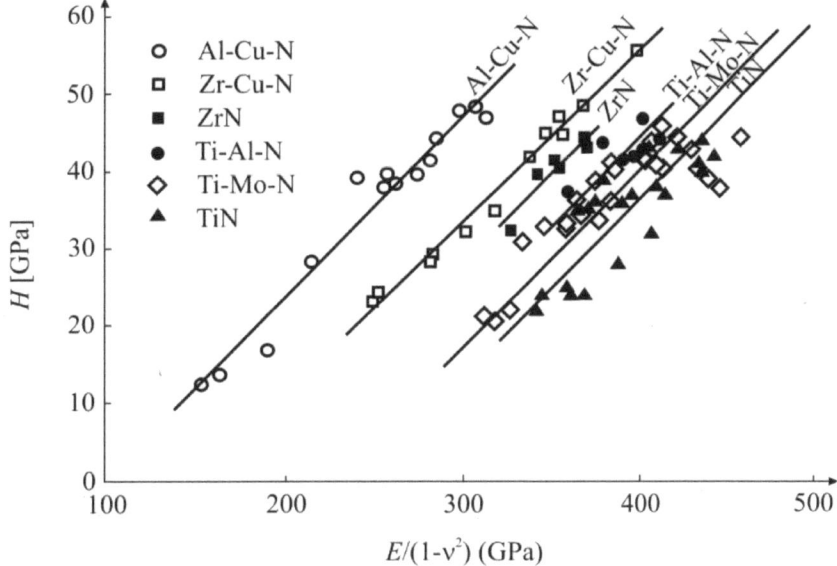

Figure 13: A dependence of H on E for a superhard coating formed by a magnetron sputtering.

On the other hand, it affects their thermal and time stability. For example in the case of ion-deposited coatings, the grain becomes smaller, a defect density increases, and a compressing stress arises. This essentially decreases the hardness, *i.e.* a stability.

Therefore, to form a nanostructured coating of a good quality, we are to optimize the coating composition, *i.e.* to select a composition, a doping element, and a quantity of doping elements for a multicomponent and a nanocomponent system, and to select a layer composition and thickness for a multilayered system.

Regulating an energy of deposited ions, optimizing a substrate temperature and composition, and controlling a reaction gas pressure in the course of deposition, we can obtain a stable coating.

8.6. EFFECT OF TEMPERATURE ON PROPERTIES OF NANOCRYSTALLINE COATINGS

As it was noted above, the high physical and mechanical properties of the nanocrystalline film depend on the nanostructure. The nanostructure, however, is a metastable phase. It means that if a film-formation temperature reaches or exceeds a crystallization one T_{cr}, a film material is crystallized. This destructs the film nanostructure and results in the formation of new crystalline phases. This is the reason, why the nanocomposite film looses its unique properties at $T > T_{cr}$. The nanostructured nanocomposite film is destroyed, and the new crystalline phases appear at T_{cr} crystallization temperature. The T_{cr} determines a thermal stability of the nanocomposite material. The crystallization temperature T_{cr} for a solid nanocomposite film is less than 1000 °C. However, $T_{cr} \approx 1000$ °C is insufficient that the nanocomposite coating will become a new material and function at high temperatures (for example, as a milling tool, or a thermal isolation of certain mechanical parts, *etc.*). Therefore, a development of the nanocomposite materials, which are thermally stable against a crystallization and oxidation resistant at above 1000 °C, is necessary. The high thermal stability of properties and the high temperature resistance to an oxidation are most attractive properties of future nanocomposite coatings. These properties strongly depend on a phase composition and the thermal stability of film-composing individual phases.

The coating oxidation is determined as a start of changes in Δm mass after thermal annealilng due to an oxide formation. The mass increases ($\Delta m > 0$), when the oxide is in a state of a solid particle. On the contrary, when a volatile oxide is formed, this leads to a loss of the mass ($\Delta m < 0$). Fig. **14** shows a mass increment as a function of an annealing temperature T.

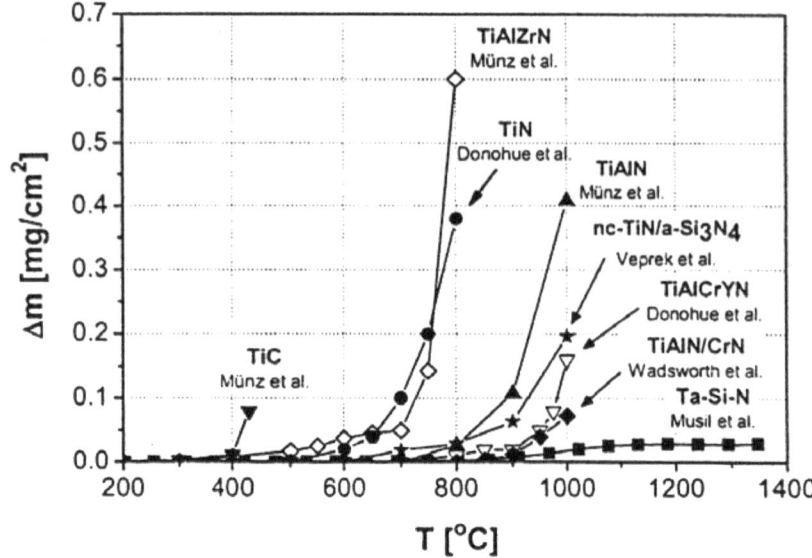

Figure 14: An oxidation resistance for certain solid coatings characterized by Δm dependence on an annealing temperature [35-37].

A temperature corresponding to the sharply increased Δm is determined as a maximum temperature T_{max}, at which a film can avoid an oxidation. The higher is T_{max}, the higher is the oxidation resistance. Types of films (a crystalline or a nanocrystalline) with the sharply increasing Δm are presented in Fig. 14. For all these films, the oxidation resistance is lower than 1000 °C, since they are composed of grains, which always have a contact with a surrounding medium through the film surface and with a substrate through the grain interface. This phenomenon essentially decreases the oxidation resistance inside the film volume and decreases its barrier capability. However, an employment of an intergranular glass-like phase allows a certain improvement. There is one way to increase the oxidation resistance of the solid-to interrupt a continuous path along the grain boundaries, starting from the coating surface through its total volume to the substrate. This can be reached in an amorphous solid film (Fig. 15).

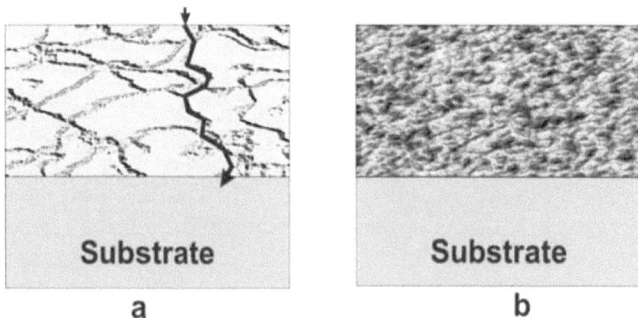

Figure 15: A schematic image of a substrate contact with an external atmosphere through a solid nanocomposite film: a-a film; b-an amorphous film [37].

Therefore, the high oxidation resistance at high temperatures can be provided by a thermal stability of both phases of the nanocomposite material: the crystallization resistance of an amorphous silicon nitride and a decomposition resistance of a metal-nitride ($MeN_x \rightarrow Me+N_g$). It can be easily reached employing a new family of $\alpha\text{-}Si_2N_4/MeN_x$ composite with a high content of $\alpha\text{-}Si_3N_4$ phase (≥ 50 vol.%). The coating is amorphous and indicates 20GPa to 40 GPa hardness, *i.e.* its hardness is adequate, which allows its application in many fields, for example, as a protecting coating of a cutting tool.

A heterophase nanocomposite coating $nc\text{-}(Ti_{1-x}Al_x)N/\alpha\text{-}Si_3N_4$ and $TiN/\alpha\text{-}Si_3N_4/TiS_2$ features a high thermal stablility, which is provided by an ideal uniform structure. Such structure is formed in the course of a "spinodal" decay and a crystallization of an amorphous nitride. The nanocomposite components are ideally mixed, therefore a generation of a crystalline phase occurs simultaneously and without the substantial increase. A conjugation of the phase interface also is high, and evidently keeps a high internal stress.

Fig. **16a, 16b** shows that a size of nc-$(Ti_{1-x}Al_x)N/\alpha$-Si_3N_4 nanocomposite crystallite remains unchangeable at 3nm level up to 1100°C. The size of TiN/α-Si_3N_4/TiS_2 nanocomposite (Fig. **16b**) increased non-essentially from 10.5 to 12nm up to 800 °C.

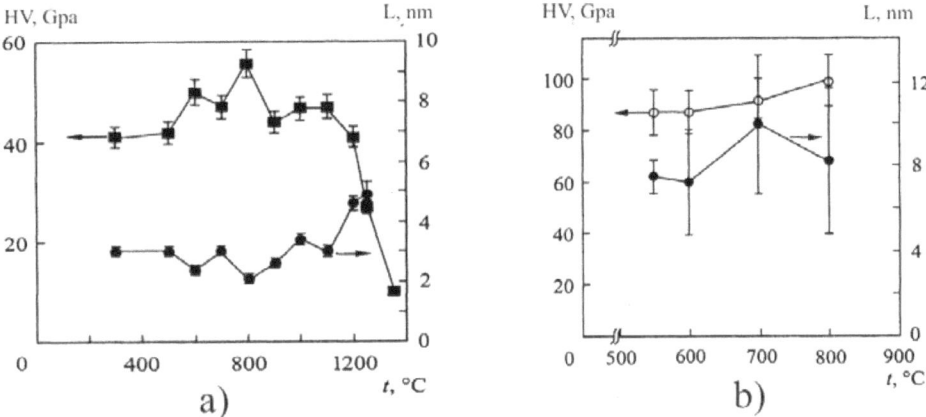

Figure 16: A change of a hardness and a crystallite size depending on a temperature of nanocrystalline coating: a-$(Ti_{1-x}Al_x)N/\alpha$-Si_3N_4; b-TiN/α-Si_3N_4/TiS_2.

One also should note that a certain hardening takes place in a system when a temperature increases. At the same time, when a grain starts to grow intensively, its properties abruptly deteriorate. The nanocomposites W-Si-N, Ta-Si-N, Ti-Si-N, Zr-Si-N with a high content of a silicon nitride phase (> 50 vol.%) are very promising, Fig. **17**.

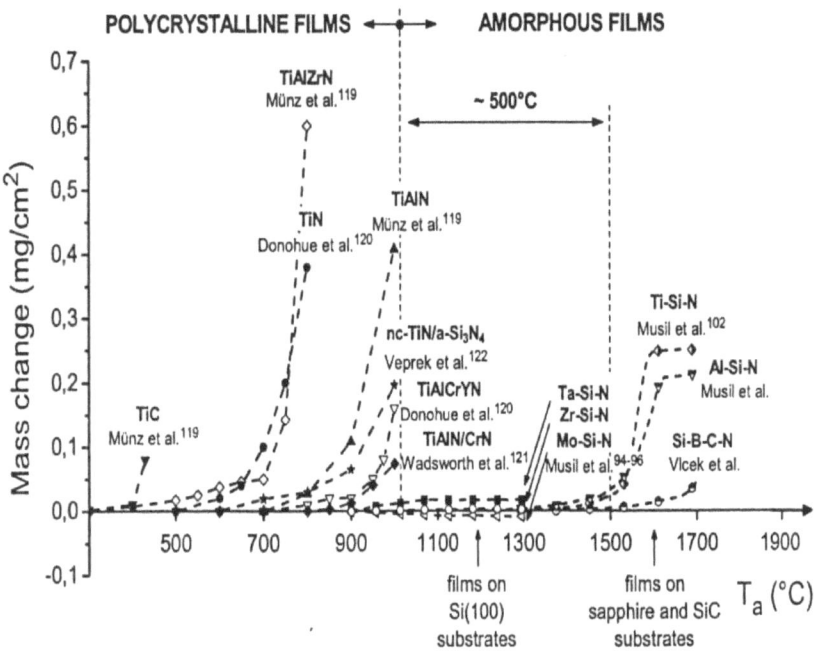

Figure 17: Dependence of mass loss on a temperature in a solid nanocomponent coating.

Ta-Si-N film with a low Si content (7at.%) is thermally stable up to 900°C. The recrystallization resulted in the formation of an intermetallic phase Ta_2N, Ta_5Si, and Ta_5Si_3. In addition, many of them demonstrate a perfect high-temperature resistance to the oxidation, which essentially exceeds 1000°C. For example in air, Zr-Si-N film does not oxide below $T = 1300°C$ [33, 37, 38].

8.7. SUPERHARD (80-100 GPA) COATINGS

Up to date, the superhardness of 70-80GPa was achieved only in a three-layer nc-TiN/a-Si_3N_4/a-$TiSi_2$ [19, 26, 33], and of 100GPa in a four-layer nc-TiN/a-Si_3N_4/a-$TiSi_2$/nc-$TiSi_2$ [19], when an oxygen content was below 0.07at.% (see Ref. [26]). Fears that such high hardness is a result of a measuring error [19] were dismissed as irrational by the results of the work [26]. In an earlier study, these estimates were compared with the data for a bulk diamond single crystal as illustrated in Fig. **18a**. It can be seen that superhard coatings have a load-invariant hardness of 100GPa that slightly decreases at a maximum load of 100mN when the indentation depth reaches about 17% of the coating thickness (3.5μm). A hardness of 91GPa under such a load was varified both by measuring the residual plastic deformation area with calibrated scanning microscopy and by computation [19]. The load-invariant hardness of the nanocomposite coating is in an excellent agreement with the bulk diamond hardness within an entire 30-70mN range and with the hardness of a single-phase nanocrystalline diamond under low load.

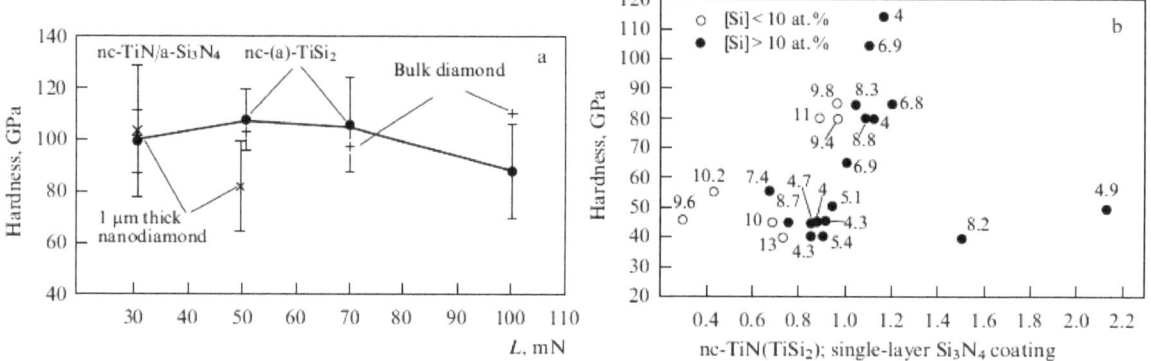

Figure 18: *(a)* Estimates of a load-independent hardness of the coating. The hardest is compared with that of a bulk diamond and a single-phase nanodiamond coating. Under 100mN load, the hardness decreases because the indentation depth is by 10% higher than the coating thickness of 3.5μm. The hardness was also measured by a calibrated SEM [26]. *(b)* A hardness of three-and four-layered nc-TiN/a-Si_3N_4/a-(nc-)$TiSi_2$ depending on the thickness of the nc-TiN coating or nc-$TiSi_2$ nanocrystalls with Si_3N_4. Arabic numerals denote a crystallite size.

A self-consistency of these data was checked by Argon and Veprek [27-29], who employed a universal binding energy relation (UBER) [29-30] and a Herzian analysis for the indentation measurements [19]. Veprek and co-workers conclude that the aforementioned high hardness values were measured correctly. It follows from the work [31] that both a high purity and an adequate composition of Si-containing phase are needed for a reproducible coating deposition. A maximum hardness is achieved when TiN nanocrystal surface is covered, at least, with one Si_3N_4 monolayer. Although the deposition reproducibility for these nanocomposites cannot be easily reached in comparison with a binary coating. The hardness exceeding 80GPa was reproduced in more than 15 coatings; a value of > 100GPa was achieved in six other cases [26].

High hardness was initially attributed to the presence of $TiSi_2$ phase. However, its strong dependence on an oxygen impurity (below 0.1 at.% content) poses a question of whether $TiSi_2$ acts as an impurity trap and prevents impurity accumulation at TiN,Si_3N_4 interface. An in-depth study is required to answer this question.

The above coatings are of interest since we need a material, the hardness of which is higher than that of a diamond. From a practical viewpoint, $TiSi_2$-containing coating has a disadvantage of the phase instability after long-term exposition in air. A deteriorated hardness, which Li *et al.* [26] observed in coatings after a few month exposition, was first attributed to a relatively high chlorine content. However, the hardness of an ultrahard nc-TiN/a-Si$_3$N$_4$/a-(nc-)TiSi$_2$ coating also decreased after 6-8-month exposition, in spite of a lower chlorine content (usually < 0.5 at.%) [19]. These researchers explained this instability by a presence of silicon titanate, which formed several phases with a different stoichiometry, such as Ti$_5$Si$_3$, TiSi, and TiSi$_2$ (the last was found to be most stable) [26, 29, 31]. However, experiments performed at a relatively low temperature (500°C-600°C) demonstrated a metaslable TiSi$_2$ phase [19], which was further transformed to a stable one only at high temperatures. Since this phase transition was kinetically confined in a small volume, TiSi$_2$ nanocomposite cxoatings were susceptible to a chemical attack for the part of polar water molecules [47-50]. This situation is reminiscent of that with a silicon fiber, in which an ideal tensile stress of 24 GPa can develop immediately after preparation. True, its hardness rapidly decreased in a moist air due to microcracking (see a monograph [32],) and fell to 0.3GPa for 24 hours.

8.8. MORPHOLOGY AND MICROSTRUCTURE

The development of a columnar morphology of a PCVD-and PVD-fabricated coating prepared at relatively low temperatures (300°C-600°C) is described by a Thornton's structure-zone diagram (see a handbook [19], p. 712). This diagram is an extension of the Movchan-Demchishin structure-zone diagram for a film, which is fabricated by gas-phase epitaxy (see [19, 26]). A proper combination of a homologous temperature $T_h = T/T_m$ (T is a deposition temperature, T_m is a melting temperature of a given material, K) and an energy transferred by ions toward a growing film surface on a deposited atom basis permits coating fabrication within a so-called "T-zone" corresponding to a columnar but a relatively dense structure.

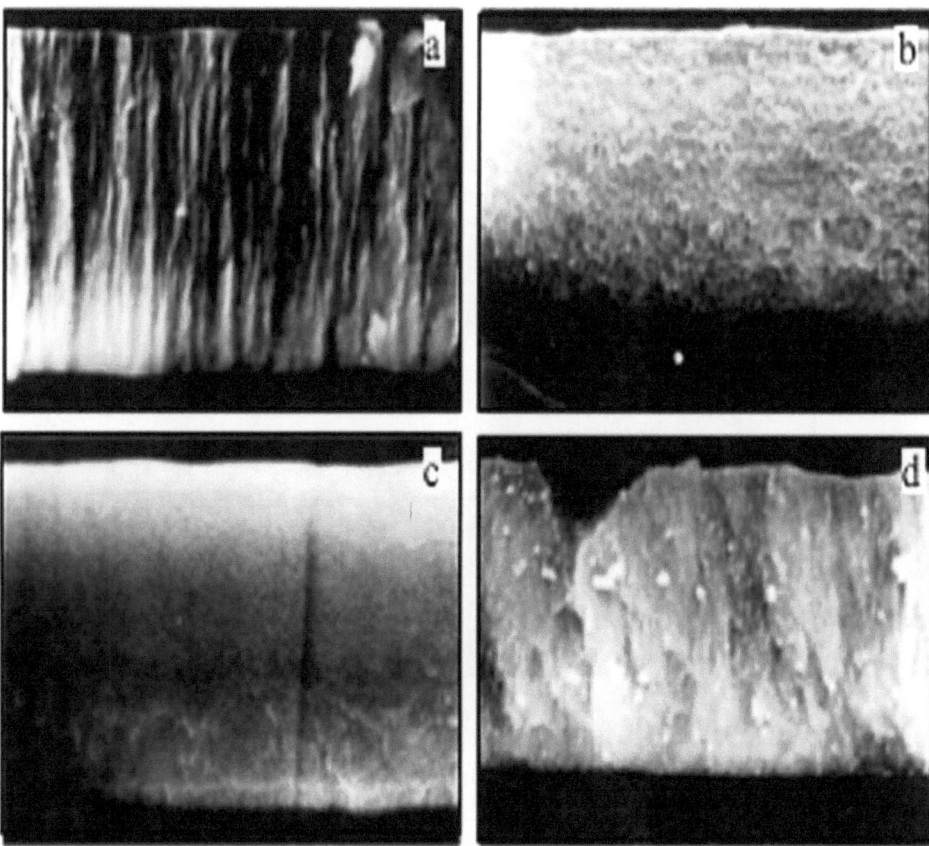

Figure 19: (*a*) A cross section through TiN coating with a columnar structure fabricated using the PCVD from TiCl$_4$, SiH$_4$ with an excess of N$_2$ and H$_2$, at 550°C; (*b*) nc-TiN and a-Si$_3$N$_4$ nanocomposite containing 5 at.% of Si and having a hardness of ca. 35 GPa; (*c*) nc-TiN/a-Si$_3$N$_4$ nanocomposite with an optimal Si content (7 to 8 at.%) and a stable nanostructure (a hardness of about 50 GPa), and (*d*) nc-TiN/a-Si$_3$N$_4$ nanocomposite with a too high content of Si (ca. 9 at.%) and a hardness of 30 GPa (taken from Ref. [19]).

The columnar structure practically disappears when a superhard nanocomposite of an optimal composition is formed. This was shown by Li *et al.* [19] as early as 1992 for the PCVD-fabricated Ti-Si-N film and later was confirmed in the works of Veprek and coworkers for another binary nanocomposite nc-MeN/α-Si$_3$N$_4$ (where Me = Ti, W, V) fabricated using the PCVD with an intense discharge [19].

Fig. **19** shows the morphological development of nc-TiN/a-Si$_3$N$_4$ after Si addition (notice that in all these coatings, a silicon is present in the form of Si$_3$N$_4$ with lacking contribution from the TiSi$_2$ phase) [26]). The TiN columnar structure (Fig. **19a**) and the surface morphology are isotropic when Si$_3$N$_4$ content is high and correspond to an optimal composition and a maximum hardness (Fig. **19c**). It should be noted that a coating deposited to Si substrate, as described in [19], cracked, whereupon a microdiagram had to be taken without surface polishing. The morphology remained unchanged after a further rise in Si$_3$N$_4$ content, but the hardness again decreased (Fig. **19d**). A similar behavior is described in [26] for nc-W$_2$N/a-Si$_3$N$_4$ and many other nanocomposites (nc-VN/a-Si$_3$N$_4$, nc-TiN/a-Si$_3$N$_4$/a-(nc-)TiSi$_2$, nc-(Al$_{1-x}$Ti$_x$,)N/a-Si$_3$N$_4$). Unfortunately, obtained results were not published due to a lack of novelty (in opinion of Prof. Veprek *et al.*). Fig. **20** illustrates a very similar development of nc-TiN/a-Si.iN4 morphology after magnetron sputtering (see Fig. **18a** and the above discussion).

Despite an isotropic morphology, which is featured by a nanocomposite with a maximum hardness deposited by magnetron sputtering, an in-depth TEM study shows a partially columnar, even if rather dense, structure [19]. It may be a consequence of relatively low nitrogen pressure, which precludes a completion of the nanostructure formation. Further studies are needed for better understanding of this problem.

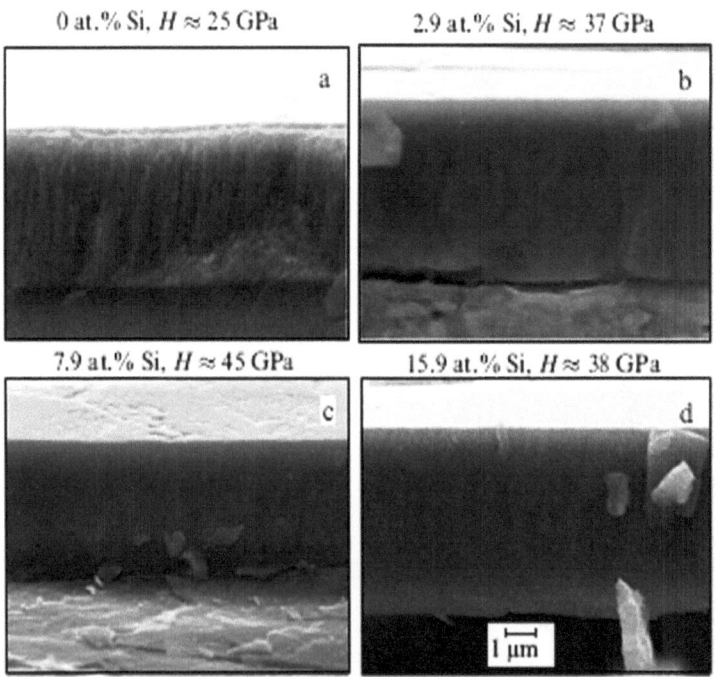

Figure 20: A development of nc-TiN/a-Si$_3$N$_4$ morphology after magnetron sputtering with a growing Si content.

The dense and isotropic nanostructure, which is developing after the formation of a stable superhard nanocomposite even within the dense T-zone, is far from an "ideal" in terms of mechanical properties by virtue of a weak intercolumnar link. Hence, the columnar morphology of the PCVD-fabricated nanocomposite needs to be eliminated altogether if a material with excellent mechanical characteristics is to be fabricated. A universal morphology dependence on the formation of a stable superhard nanocomposite is confirmed by earlier reports dealing with an identical isotropic structure in nc-TiN, a-BN, and nc-TiN/a-BN/a-TiB$_2$ coatings. Certain research groups reproduced these data and thereby confirmed them to a degree [19]. However, results of a scanning electron microscopy (SEM) presented in these publications indicate that the coatings retain a well-apparent columnar morphology even after silicon addition [26]. This can probably be accounted for by the fact that all these authors used a lower partial pressure of a nitrogen than Procházka et al. [19].

The authors of [19] also discuss the mechanisms governing the aforementioned morphological changes but totally disregard the phase segregation concept formulated a few years ago [19]. For example, Patscheider et al. [33, 34] postulate a certain type of nonspecific "renucleation" of TiN grains following "partial deposition of Si_3N_4 onto TiN". The authors do not even refer to the thermodynamic and kinetic considerations presented, nor do they go into the details of the renucleation mechanism that regulates epy morphological development. However, SEM micrograms presented in [19] show that Si-containing coatings (12 at.%) with a maximum hardness around 40GPa [26, 33] obviously retained a well-apparent columnar structure [19].

Hu et al. [26] also revealed a similar tendency; in their study, a rise in Si content from 0 (pure TiN) to 5 at.% did not result in disappearance of the columnar morphology. The present authors conclude that speculative arguments presented in [26] fully ignore the thermodynamic and kinetic data reported over 10 years ago by Veprek et al. [19]. The results of Hu and coworkers are easy to understand, taking into account the hardness values presented in their paper. The maximum hardness (ca. 36GPa) was obtained only for a coating with 4at.% of Si deposited at room temperature; those prepared at 400°C had their maximum hardness reduced to 28 GPa. This means that the hardness enhancement in the coating deposited at room temperature is first and foremost a result of a bombardment with high-energy particles, which are always present in the magnetron sputtering at a low gas pressure due to a reflection of primary ions from a target. A base pressure employed by Hu et al. was 10^{-6} mbar, and a partial nitrogen pressure was 2×10^{-4} mbar. The authors interpreted this fact as suggesting the presence of a silicon, not only in the form of Si_3N_4 but also as Si (true, the critically disposed reader will notice that the Si 2p XPS signal ascribed by the researchers to the elemental Si might just as well originate from $TiSi_2$). Collectively, these results indicate that coatings of Hu et al. and Patscheider et al. were deposited under improper conditions that prevented the formation of a stable fully segregated nanostructure during the film growth and accounted for the prevailing columnar morphology, even if less pronounced.

Many authors reporting a different type of a behavior of the hardness, the morphology, and other properties suggest an alternative interpretation. However, it is difficult to analyze these works for the lack of the comprehensive characteristics of deposition conditions and/or test samples and due to a necessity of excluding injurious effects of impurities. By way of example, Jiang et al. [26] prepared a series of Ti-Si-N coatings at room temperature by an unbalanced reactive magnetron sputtering at an overall (Ar-N_2) pressure of 2.6×10^{-3} mbar but did not specify the partial nitrogen pressure. Hence, no wonder that only a part of an augmented silicon amount in the coatings was involved in the formation of the Si_3N_4 type linkage but the remaining Si "did not participate in the reaction", while the columnar morphology was preserved (see Ref. [26]). A small thickness of the coatings (within 1 μm) [26] necessitated a hardness measurement at a low pressure and a maximum indentation depth (580 nm). The results of these measurements need thorough verification, bearing in mind a possible indentation size effect (ISE), which is able to adulterate them. The authors of [26] observed the hardness enhancement to 35GPa when Si content ranged 8-10 at.%. However, a presence of an oxygen impurity (5-8 at.%) [26] makes this work of a little value for understanding a system of this kind. Such criticism equally applies to many other studies where a coating was deposited ". without applying a substrate bias voltage or a heating, for an evaluation of the added silicon effect alone." [26]. A rise in the deposition temperature to 300°C and an application of -100V bias voltage [19] led to the hardness enhancement unrelated to a bombardment with high-energy ions or a correlation between the hardness and the compressive stress.

8.9. MECHANICAL PROPERTIES OF SUPERHARD NANOCOMPOSITES

An understanding of the extraordinary mechanical properties of a nanocomposite material was initially based on an absence of a dislocation activity in a small (a few nanometer) crystallite, a very low stress level for a small (< 1 nm) nanocrack. and a resulting necessity of a high stress for the crack initiation and propagation in a system lacking a sliding along a grain interface [19, 26, 28, 30, 40-44]. A well-defined interface between nanocrystals and Si_3N_4 phase are attributed to a thermo-dynamically governed phase segregation. This situation is critically different from the nanocrystalline material, which is fabricated by means of a consolidation, when the grain interface sliding is impeded [19]. An important contribution brought about by Argon's studies [27, 28] provided a significantly deeper insight into the problem. It was shown that the mechanical properties of the nanocrystallite were easy to explain in terms of the conventional fracture mechanics, taking an account of their nanometer scale and an amorphous covalent transition silicon nitride one monolayer thick, when these materials had no cracks. This situation is due to the formation of a stable nanostructure by a self-organization during a spinodal phase segregation. Our discussion is confined to a brief summary of earlier publications on this problem.

Plastic deformation of a ductile material like a metal proceeds *via* a dislocation activity or shift transformations, such as a deformation twin or a martensitic transformation. Plastic flow in glasslike solids, for example, amorphous metals, occurs as shift transformations return to bulk ferrite elements where they are 3-4 nm in size [19]. A recently developed molecular-dynamic Si-glass model implies that the localized plastic deformation in a covalent amorphous solid is triggered by a simultaneous collective movement of 5-6 atoms. The clusters, which are redistributed clusters under the plastic deformation, contain 100-500 atoms [19]. The plastic deformation evolving by any of these mechanisms requires a finite activation volume [26]. This situation resembles that of a structure-phase transformation resulting from imposing a pressure to a nanometer-sized single crystal, where a transition pressure increases with a decreasing crystal size. A reverse transition after unloading is characterized by a hysteresis-like enhancement upon a decrease in the crystallite size [19, 28]. The plastic deformation is utterly impossible within the nanometer-sized crystallite and a single Si_3N_4 monolayer or a BN 'glue' in a stable super hard nanocomposite, the strength of which might afterwards reach an "ideal" level.

Such an ideal strength of a perfect solid (a crystal or a glass) is characterized by a decohesion energy σ_c $\sim(E\gamma_s/a_0)^{0.5}$ (E is a Young modulus, y_s is a surface energy, and u_l is an interatomic bond length [26]. Calculations based on this relation give an ideal solid strength on the order of 20-50GPa, which is close to its observed value for a freshly prepared glass fiber. Indeed, a tensile stress in a freshly prepared SiO_2 fiberglass amounts to 24GPa or 25% of the Young modulus [26]. An elastic stress for an ideal solid matter recovery can be as high as 20% (see monograph [19], Ref. [26] and references cited therein).

Based on a universal binding energy relation (UBER) [28], Prof. Argon and Prof. Veprek [29] assessed an ideal adhesive strength of nc-TiN/a-Si_3N_4 as 46GPa and compared it with a tensile stress (33 GPa) at a periphery of a contact between a coating and an indenter under an application of a maximum (70 mN) load. A Herzian analysis applied to measuring tensile stress in a glass was used in an earlier work by Argon *et al.* [28]. This procedure is employed to determine the tensile (radial) stress for a crack-free indentation in a variety of super-and ultrahard nanocomposites [19]. The radial tensile stress in such indentation represents the lower limit of the tensile stress in a nanocomposite and actually reaches an ideal value calculated for this material. The detailed Herzian analysis of the experimental indentation curves for the nanocomposite and a diamond confirms a self-consistency of the results of these measurements and a high hardness value obtained for these materials [26]. The results of a numerical simulation by a finite element method (FEM) also confirms the experimental data [33]. Other details can be found in the aforecited publications.

A high "crack toughness" of the materials under consideration is frequently applied because an indentation at a large load (*e.g.*, 1000mN) does not cause the defect formation in a 6-10-μm thick coating on a mild steel substrate with 20% stress [19] (for a confirmation of a reproducibility of these results). A thinner coating shows only the circular (not radial) Herzian cracks, which seem to originate from the corners of the indentation region. These examples demonstrate an importance of very high resistance for a brittle cracking in these materials. It does not mean, however, that they must just as well have the high crack toughness described quantitatively by the stress intensity factor K_I or the energy release rate. For a simple case of a flat crack of a size $2a$, the stress intensity factor is $K_I = \sigma(\pi a)^{0.5}$, where σ is a stress necessary for a $2a$-sized crack propagation. The value of K_I can be found by the indentation technique if the radial crack arises in the indentation corner and its length is much smaller than a sample thickness. This condition can hardly be fulfilled in a coating. Nevertheless, an absence of the radial crack in many indentation experiments (such as reported by Prof. Veprek and some other researchers) in a combination with the high tensile stress is a characteristic feature of a material having a crack initiation threshold [26].

Apart from the high brittle cracking resistance, the nanocomposites shows a high potential for an elasticity recovery (up to 94%) after imposing a load of 70 mN, when a load-independent hardness exceeds 100GPa [19] and an elasticity limit after imposing the load of 1000GPa is greater than 10%. For example, no cracks develop when a 6-μm thick coating is pressed by 1000mN into a mild steel substrate ($H_v \sim 1.8$GPa) to a depth of 2μm. A hardness of the composite involving the coating and the substrate remains equal to 40GPa, and an unloading results in a 10% recovery of the coating, despite the fact that it is practically held by an excessively plastically deformed substrate [19]. An absence of cracks even beneath the coating surface was thoroughly verified by the Herzian analysis and a load curve evaluation [26].

As shown by Prof. Veprek et al. [19], all these properties are consistent with a defect-free nature of the material. Imagine a glass of a perfect (defect-free) structure. A stress-related change in an internal energy roughly corresponds to a change in an interatomic binding energy integrated over the entire sample. A stress resistance, i.e. a restoration (recovery) force acting to minimize the system's energy (an interatomic distance and a bond angle being brought to their equilibrium values) has been considered in [19]. Notice that the restoring force is the first derivative of the internal energy in the presence of the stress. The first derivative of a tensile strength for an equilibrium stress is the Young modulus, i.e. $E = (d^2U/d\varepsilon)_0$ (ε is an elasitcity coefficient). It should be mentioned that the elasticity modulus grows after the compression ($\varepsilon < 0$) and decreases after stretching ($\varepsilon > 0$). A maximum tensile strength matches the decohesion force σ_m achieved at the maximum recovery stress (the elasticity limit) $\varepsilon_m \sim 15-20\%$. when the interatomic covalent bond breaks down under an effect of the stress [19]. The entire interval $0 < \varepsilon < \varepsilon_m$ corresponds to the elastic strain but is not linear due to the stress dependence of the elastic modulus.

This line of reasoning leads to a conclusion that the extraordinary mechanical properties of the superhard stable nanocomposite are easy to explain in terms of a conventional fracture mechanic for a material that is free of cracks. A freshly prepared silicon fiber and a whisker (a filamentous single crystal of a few nanometer thickness) has practically no cracks and the stress in them is close to ideal. As it follows from Fig. **19** [19], stress in these materials is equivalent to that in a stable nanocomposite.

Future studies should be designed to provide a deeper insight into a nature of the nonlinear elastic response and a mechanism of the plastic deformation in these materials. For example, a bulk modulus B in an area under the indenter for a coating with 50-60GPa hardness increases with a rise of a pressure P as $B(P)$ a $B(0) + 5P$ where $B(0)$ is the bulk modulus in an absence of the pressure. A factor of proportionality of 5 measured by a high-resolution X-ray diffraction method agrees with a predicted value based on the UBER [28, 29]. For this reason, the bulk modulus measured by the high-resolution X-ray diffraction method [26] for the coatings prepared by Veprek et al. $B(0) = 295 \pm 15$GPa, increase with a rise in a hardness up to 545-595GPa (note that the pressure under the indenter is roughly equal to H= L/A_c, where L is the indentation load, and A_c is the coating area). This makes the routine finite element method (FEM) unfit to resolve a problem of a strong nonlinearity even though a major part of the elastic reaction after unloading occurs in a region lying far apart from the intender, where the pressure grows insignificantly, while the stress value to be used in a computation is rather high. Such a situation arises when a mechanism of the plastic deformation operates in an amorphous material [19]. The FEM-assisted simulation of the stress yields very interesting data relevant to the problem under discussion [19, 26].

Fig. **21** presents the results of hardness measurements obtained with the help of the nanoindenter for a superhard (or ultrahard) coating deposited onto different substrates (see also Fig. **18a** and **18b**). These data are reported by Prof. Veprek and coworkers in Refs. [26] and [19] (see also Ref. [10]); they are discussed there to show how the hardness was measured. Fig. **21b, 21c** displays no radial cracks in the SEM image in a diagonal direction. According to the authors of Refs [19, 26], the absence of radial cracks in many indentation experiments, coupled with the high tensile stress, is an intrinsic feature of a material having the crack initiation threshold [19].

Figure 21: Scanning electron micrograms of an indentation site under a load of 1000 mN in a superhard coating: (*a*) 3.5-μm thick coating (a load-invariant hardness $H_v \sim 105$ GPa, Fig. 18a); the coating is pressed deep into a mild steel substrate; (*b*) 10.7-μm thick coating (the load-invariant hardness $H_v \sim 60$ GPa); (*c*) 6.1-μm thick ultrahard coating (the load-invariant thickness $H_{0.005} \sim 100$ GPa, a coating/substrate composite hardness is around 38GPa). Note an absence of radial cracks in a diagonal direction in the images (*b*) and (*c*) [19].

8.10. PROSPECTS FOR THE FIELD

A further research in a field of the nanostructured film will be carried out along the following lines: (*1*) the formation of a film with a controllable grain size in a range from 1 to 10nm for an elucidation of a relationship between its properties and the dimensional parameters, and the fabrication of a new improved coating with unique physical and functional characteristics; (*2*) a nanocrystalization from an amorphous phase; (*3*) a transfer of an electron charge between nanograins of different chemical compositions and a Fermi energy; (*4*) a preparation of a new protective coating with an oxidation resistance higher than 2000 °C, and (*5*) a development of a new system for a deposition of a nanostructured coating by physical methods. It may be expected that the thin nanostructured film will be used in the nearest future as an experimental model for the fabrication of a nanostructured bulk material with tailored properties.

To date, TiN, TiNCr, and TiZrN coatings and TiN/CrN, MoNTiAlN multilayer coatings, and some other materials have found a variety of applications in industry. An implantation of certain additional components (*e.g.*, Si or B) into TiN film can significantly improve its physical and mechanical properties and thereby widen a spectrum of its application. A use of the thin coating is fairly well described in a recently published review [6] dealing with an application of a multifunctional nanostructured film.

Table **2** presents a few examples of the application of a tribological superhard nanostructured coating composed of immiscible phases (see also references [33, 37-45, 47] in this review).

It is shown that doping a classical TiN coating with Si, B, Al, or Cr ensures a combination of the high hardness and a durability with a sufficiently low friction coefficient [26, 28]. These multi-component nanostructured coatings and films may be used to protect a surface of different tools and instruments [49-50], such as a cutting and stamping tool, a casting roller, parts of an aircraft engine, a gas-driven turbine and a compressor, a sliding bearing, and an extrusion nozzle for a glass and mineral fiber, against a simultaneous action of a high temperature, a mechanical wear, and an aggressive medium [6].

Table 2: Examples of tribological superhard nanostructured films comprising immiscible phases (see review [6] and references cited therein).

Coating	Target composition	Film phase composition	Literature (references taken from review [6])
Ti-B-C-N	TiB + Ti or TiB$_2$+TiC	nc-TiC + nc-TiB$_2$ + a-BN	[37]
Ti-B-N	TiB$_2$+TiN or TiB$_2$+Ti	nc-TiN + nc-TiB$_2$ (or a-TiB$_2$) + a-BN	[33, 36. 38]
Ti-Si-N	Ti$_5$Si$_3$+Ti	nc-TiN + nc-TiSi$_2$ + a-Si$_3$N$_4$	[39]
	a*	nc-TiN + a-Si$_3$N$_4$	[40]
Ti-Cr-B-N	TiB+Ti$_9$Cr$_4$B+Cr$_2$Ti	nc-TiN +CrB2 + a-BN +a-TiB$_2$	[33]
Ti-C-B	TiC-TiB;	nc-TiB$_2$ + TiC + a-B$_4$C	
Ti-Si-B-N	TiB$_2$ + Ti$_5$Si$_3$ + Si or TiB$_2$+ Si	nc-TiB$_2$ + TiN + nc-TiSi$_2$ + a-Si$_3$N$_4$	[35. 36]
Ti-Si-C-N	TiC + Ti$_5$Si$_3$ or TiC + Ti$_5$SiC$_2$ + TiSi$_2$ + SiC	nc-TiC + nc-Ti$_2$SiC$_2$ + nc-TiSi$_2$ + SiC + a-Si$_3$N$_4$	[36.42)
Ti-Al-B-N	TiAlBN	nc-TiB$_2$ + nc-(Ti$_x$Al$_{1-x}$)N + a-BN + a-AlN	[1]
Ti-Al-C-N	TiAl + TiC	nc-TiC + nc-(Ti$_x$Al$_{1-x}$)N + a-AlN	[36]
W-C	W + laser ablation of C	nc-WC + a-C	[43]
W-Si-N	b*	nc-W$_2$N + a-Si$_3$N$_4$	[44]
Ti-C	TiC$_{0.5}$	nc-TiC + a-C	[45]
Ti-Al-Si-N	TiN + TiAl	nc-(Ti$_x$Al$_{1-x}$)N + a-Si$_3$N$_4$	[46]
Cr-Si-N	Cr + Si	nc-CrN+ a-Si$_3$N$_4$	[47]
Cr-B-N	CrB$_2$	nc-CrB$_2$ + a-BN	[33]

Notes: Most films were prepared by a magnetron sputtering of targets in an atmosphere of Ar or N$_2$/Ar; *a**: films obtained by a chemical vapor deposition (CVD); *b**: films fabricated by thermal annealing of an amorphous W± Si ±N layer.

Other promising fields of an application of these coatings include a magnetic optics, an electronic engineering, and a medicine (a DNA synthesis). Moreover, these coatings are needed to develop a new generation of a biocompatible material, *e.g.*, an orthopedic and a dental implant, a material for a cranial and a maxillofacial surgery, and a fixation of a cervical and a lumbar spine, as described at greater length in review articles [19, 26].

QUESTIONS FOR CONTROL

1. Under what conditions a nanocrystalline film is formed?
2. What is an effect of an energy transferred by ions in a magnetron sputtering?
3. What is a role of energy in a reactive magnetron sputtering?
4. What are formation features of a nanocrystalline film and a coating?
5. What is a nanocomposite coating?
6. What groups of nanocomposite coatings exist, and how are they classified?
7. What is an increased hardness of a nanocomposite coating related to?
8. How does a coating structure affects thermal properties?

REFERENCES

[1] Andreev AA, Sablev VP, Shulaev VM, Grigoriev SN. Vacuum-Arc Devices and Coatings. Kharkov: NNTs KhFTI 2005.

[2] Azarenkov NA, Beresnev VM, Pogrebnjak AD. Structure and Properties of Protecvite Coatings and Modified Layers of Materials. Kharkov: KhNU 2007; 565.

[3] Andrievskii RA. Formation and Properties of Nanocrystalline Refractory Compounds. Russ Chem Rev 1994; 63: 431-448.

[4] Beresnev VM, Pogrebnjak AD, Azarenkov NA, *et al.* Nanocrystalline and Nanocomposite Coatings, Structure, Properties. Phys Surf Eng 2007; 6: 4-27.

[5] Beresnev VM, Pogrebnjak AD, *et al.* Structure, Properties, and Formation of Solid Nanocrystalline Coatings Deposited Using Several Methods. Uspehi Fiziki Metallov 2007; 8: 171-246.

[6] Levashov EA, Shtanskii DV. Multifunctional Nanostructured Films. Russ. Chem. Rev. 2007; 76: 502-509.

[7] Reshetniak EN, Strelnitskii VE. Synthesis of nanostructured Films: Achievements and Perspectives. Problems of Atomic Science and Technology 2008; 2: 119-130.

[8] Drobyshevskaia AA, Davydov IV, Fursova EV, Beresnev V.M. Nanocomposite Coatings Based on Nitrides of Transition Metals. Phys Surf Eng 2008; 5: 93-98.

[9] Pogrebnjak AD, Shpak AP, Azarenkov NA, Beresnev VM. Structure and Properties of Hard and Superhard Nanocomposite Coatings. Physics Uspekhi. 2009; 52(1): 29-54.

[10] Azarenkov NA, Beresnev VM, Pogrebnjak AD. Structure and Properties of Protecvite Coatings and Modified Layers of Materials. Kharkov: KhNU 2007; 565.

[11] Vudraf D, Delchar T. Sovremennye Metody Issledovaniia Poverkhnosti. Eng.Transl. Moscow: MIR 1989.

[12] Movchan BA, Demchishin AV. Investigation of Structure and Properties of Vacuum-Deposited Thick Nickel, Titanium, Tungsten, Aluminum Oxide, and Zirconium Dioxide Coatings. The Physics of Metals and Metallography 1969; 28: 23-30.

[13] Shulaev VM, Andreev AA, Gorban VF, Stolbovoi VA. Comparison of Characteristics of Vavuum-Arc Nanostructured TiN Coatings Deposited by Application of High-Voltage Pulses to Substrate. Phys Surf Eng 2007; 6: 94-98.

[14] Korotaev AD, Moshkov VYu, *et al.* Nanostructured and Nanocomposite Superhard Coatings. Physical Mesomechanics 2005; 8: 103-116.

[15] Andrievskii RA. Nanomaterials: Concepts and Modern Problems. Rus Chem Jour 2002; XLVI: 50-56.

[16] Beresnev VM, Tolok VT, Shvets OM, *et al.* Micro-Nanolayered Coatings Formed by Vacuum-Arc Deposition Using HF Discharge. Phys Surf Eng 2006; 4: 93-97.

[17] Veprek S, Karankova P, Maritza G, Veprek-Heijman GJ. Possible role of oxygen impurties in degradation of nc-TiN/a-Si_3N_4 nanocompositete. J Vac Sci Technol B 2005; 23: L17-L21

[18] Loktev YuD. Nanostructured Coatings for High-Capacity Tools. Struzhka magazine 2004; 2(5): 12-17.

[19] Turbin PB, Beresnev VM, Shvets OM. Nanocrystalline Coatings Formed by Vacuum-Arc Method Using HF Voltage. Phys Surf Eng 2006; 4: 198-202.

[20] Kunchenko YuV, Kunchenko VV, Nekliudov IM, *et al.* Layered Ti-Cr-N Coatings Formed by Vacuum-Arc Deposition Method. Problems of Atomic Science and Technology 2007; 2(90): 203-214.

[21] Ragulia AV, Skorokhod VV. Konsolidirovannye Nano Strukturnye Materialy. Kiev: Naukova Dumka 2007.

[22] Veprek S, Argon AS. Towards the understanding of the mechanical properties of super-and ultrahard nanocomposites. J Vac Sci Technol 2002; 20: 650-664

[23] Kukushkin SA, Slezov VD. Dispersnye Sistemy na Poverkhnosti Tviordykh Tel (Evolutsionnyi Podkhod) Mekhanizmy Obrazovaniia Tonkikh Plionok. St.-Pererburg: Nauka 1996.

[24] Protsenko IYu. Tekhnologiia ta Fizika Tonkikh Metalevykh Plivok. Ukraine, Sumy: SumDU 2000.

[25] Kunchenko VV, Kunchenko YuV, Kartmazov GN, *et al.* Nanostructured Superhard nc-TiN/a-Si_3N_4 Coatings Formed by Vacuum-Arc Deposition Method. Problems of Atomic Science and Technology 2006; 4(89): 185-190.

[26] Pogrebnjak AD, Danilionok MM, Uglov VV. *et al.* Nanocomposite protective coatings based on Ti-N-Cr/Ni-Cr-B-Si-Fe, their structure and properties. Vacuum 2009; 83(1): S235-S239.

[27] Argon AS, Veprek S, In: Proc. 22th Riso Int. Symp. on Materials Science: Science of Metastable and Nanocrystalline Alloys, Structure, Properties and Modeling, ed. by A. R. Dinesen, M. Eldrup, D. Juul Jensen etal., Riso Nat. Laboratory, Roskilde, Denmark 2001.

[28] Valiev RZ, Aleksandrov IV. Nanostructured Materials Formed by Intensive Plastic Deformation. Moscow: Logos 2000.

[29] Pogrebnjak AD, Beresnev VM, Il'yashenko MV, *et al.* Features of Structure and Properties of Solid Ti-Al-N and Superhard Ti-Si-N Nanocomposite Coatings Deposited by CVD in HF Discharge. Phys Surf Eng 2008; 6: 221-227.

[30] Niederhofer A., Bolom T, Nesladek P, et al. The role of percolation threshold for the control of the hardness and thermal stability of super-and ultrahard nanocomposites. Surf Coat Technol 2001; 146-147: 183-188.

[31] Karvankova P, Veprek-Heijman MG, Zindulka O, et al. Superhard nc-TiN/a-BN and nc-TiN/a-TiBx/a-BN coatings prepared by plasma CVD and PVD: a comparative study of their properties. Surf Coat Techn 2003; 163-164: 149-156.

[32] Sobol OV. Factors Conditioning Fromation of Amorphous-Like and Nanocrystalline Structured State in Ion-Plasma Deposited Coatings. Physical Surface Engineering 2008; 6: 134-141.

[33] Musil J, Barosh P, Zeman P. Hard nanocomposite coatings. Present status and trends. ch.1, in book Ed. R.Wei Plasma surface engineering and its practical application. USA. Research Singpost Publisher 2007.

[34] Musil J. Hard and Superhard Nanocomposite Coatings. Surf Coat Technol 2000; 125:322-330.

[35] Suna J, Musil J, Ondok V et al. Enchanced Hardness in sputtered Zr-Ni-N films. Surf Coat Technol 2006; 200: 6293-6297.

[36] Zeman P, Musil J, Daniel R. High-temperature oxidation resistance of Ta-Si-N films with a high Si content. Coatings. Surf Coat Technol 2006; 200:4091-4096.

[37] Musil J, Visek J, Zeman P. Hard amorpheus nanocomposite coatinds with oxidation resistance above 1000^0C. Adv Appl Ceramics 2008; 107: 148-154.

[38] Pogrebnjak AD, Sobol OV, Beresnev VM, et al. Phase Composition, Thermal Stability, Physical and Mechanical Properties of Superhard On Base Zr-Ti-Si-N Nanocomposite Coatings Nanostructured Materials and Nanotechnology IV: Ceramic Engineering and Science Proceedings 2010; 31(7): 127-138.

[39] Sung J, Lin J. Diamond Nanotechnology. Syntheses and Applications. Pan Stanford Publishing Pte. Ltd. 2010.

[40] Greim J, Schwetz KA. Boron Carbide, Boron Nitride, and Metal Borides, in Ullmann's Encyclopedia of Industrial Chemistry. Wiley-VCH: Weinheim 2005.

[41] Dubrovinskaia N, Dubrovinsky L, Solozhenko VL. Comment on "Synthesis of Ultra-Incompressible Superhard Rhenium Diboride at Ambient Pressure. Science 2007; 318: 1550c.

[42] Levine JB, Tolbert SH. Kaner RB. Advancements in the Search for Superhard Ultra-Incompressible Metal Borides. Advan Funct Mater 2009; 19: 3519.

[43] Solozhenko VL. Ultimate Metastable Solubility of Boron in Diamond: Synthesis of Superhard Diamondlike BC5. Phys Rev Lett 2009; 102: 015506.

[44] Levine JB, Nguyen SL, Rasool HI, Wright JA, Brown SE. Preparation and Properties of Metallic, Superhard Rhenium Diboride Crystals. J Am Chem Soc 2008; 130 (50): 16953.

[45] Levine, JB, Betts JB, Garrett JD, Guo SQ, Eng JT. Full elastic tensor of a crystal of the superhard compound ReB2. Acta Materialia 2010; 58: 1530.

[46] Beresnev VM, Pogrebnyak AD, Turbin PV, et al. Tribotechnical and Mechanical Properties of Ti-Al-N Nanocomposite Coatings Deposited by the Ion-Plasma Method. Journal of Friction and Wear 2010; 31 (5): 349-355.

[47] Beresnev VM, Sobol' OV, Pogrebnjak AD, et al. Thermal stability of the phase composition, structure, and stressed state of ion-plasma condensates in the Zr-Ti-Si-N system. Tech Phys 2010; 55 (6): 871-873.

[48] Pogrebnyak AD, Danilenok MM, Drobyshevskaya AA, et al. Investigation of the structure and physicochemical properties of combined nanocomposite coatings based on Ti-N-Cr/Ni-Cr-B-Si-Fe. Russian Physics Journal 2009; 52 (12): 1317-1324.

[49] Pogrebnjak AD, Lozovan AA, Kirik GV, et al. Structure and Properties of nanocomposite, hybrid and polymers coatings,Publ. House URSS, Moscow, 2011, 344.

[50] Azarenkov NA, Beresnev VM, Pogrebnjak AD, et al. Fundamentals of Fabricated Nanostructured Coatings, Nanomaterials and Their Properties, Publ. House URSS, Moscow 2012; 352.

CHAPTER 9

Application of Nanomaterials in Engineering

Abstract: This Chapter presents results demonstrating an application of a nanocrystalline coating in industry. First of all, it considers an application of a nanostructured coating for increasing servicing life of a tool and for the fabrication of site of various devices, for example, a sensor. An application of a nanomaterial in a biotechnology (including an artificial collagen), a nanofiltering for a water purification are considered.

Keywords: Industry, biosensor, bionanotechnology, CNT, cancer cell, CWCNT-polymer.

9.1. NANOCRYSTALLINE COATINGS IN INDUSTRY

Today, only several types of a nanomaterial is applied in practice. Among them are: a nanostructured nickel foil, a magnetically soft (plastic) alloy "Finemet", a multilayered semiconducting heterostructure, a superhard nitride film, *etc.* [1-3]. A coating, which can increase a servicing ability of a cutting tool, a wear resistance of a machine site, and a tool [4-16, 51-57], gains the most practical application in industry. Table **1** presents certain applications of the nanocrystalline coating for a cutting tool in comparison to a standard TiN coating. A nanostructured multilayered film formed on a cubic BN, C_3N_4, TiC, TiN, Ti(Al,N) base has a very high or an ultrahigh hardness (reaching 70GPa). It also gains a good recommendation for a sliding friction behavior. There is a number of films working under a shock wear. The development of a superhard nanostructured nitride film is reported in [2, 13, 17, 58-62]. Also, the good tribotechnical properties of an amorphous or a nanostructured film of carbon and carbon nitride are reported in [14]. The work [15] reports about TiC, TiN, and TiCN. A multiphase nanostructured coating of 20GPa hardness and 0.05 wear-sliding coefficient based on TiB_2-MoS_2 fabricated on a steel substrate [8] are offered as a self-lubricating coating for a space engineering. A metallic nanopowder is added to a motor oil for recovering of a wearing surface [16]. Fig. **1** illustrates an effect of an aluminum content in TiAlN coating on the mechanical and tribotechnical properties.

Table 1: Results of industrial tests performed for cutting tools coated with a nanostructured and a nanocomposite coating.

N	Coatings	Treated Material	Tools	Coefficient of Increase	Treatment
1	n-TiN_x/CrN_x	(NiCr)70,B,Si,Fe	WC-(Co8%)	7.0	Rough
2	n-TiN_x/CrN_x	(NiCr)70,B,Si,Fe	WC-(Co8%)	3.5	Scum
3	n-TiN_x/CrN_x	Ni,Co,Fe,Si	WC-(Co8%)	3.0	Rough
4	nc-TiN/αSi_3N_4	Marbled limestone	5Mo6V1.2WFe	2.0-3.0	Drilling
5	$Ti_{1-x}Al_xN$	12Cr5Ni Fe	TiC85%Co6%	3.0-4.0	Rough
6	n-$Ti_xAl_{1-x}N$	38Cr3NiFe	WC-(Co8%)		Rough

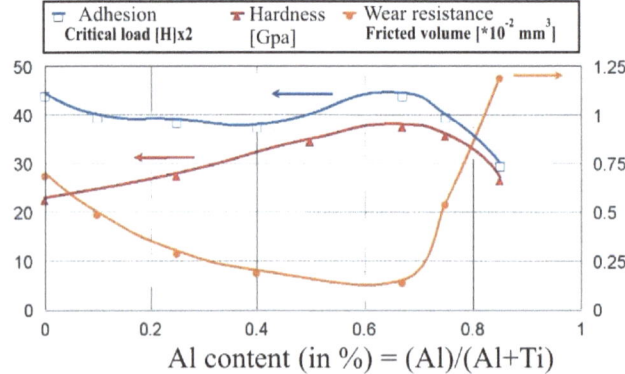

Figure 1: The mechanical and tribotechnical properties of TiAlN coating [12].

In works [18-20], an application of the nanostructure coating in a navigation, a broadband system of an electromagnetic protection, a fabrication of a high temperature fuel element, *etc.* are considered. An evaluation of a serviceability of the nanostructured coating based on Ti-Si-C system, which is deposited to a shoulder-blade of a compressor gas turbine motor (*GTM*) (a many-time heating with an subsequent air-cooling) demonstrates a higher thermal resistance than a multilayered one.

9.2. APPLICATION OF NANOSTRUCTURE AS DEVICE ELEMENT

A step to new fabrication methods, especially, a tool fabrication using a "bottom-to-top" atom-by-atom or a molecule-to-molecule formation seems to be realized in the nearest ten years. However in this case, we shall speak about the nearest future for certain nanomaterials or nanostructures. Three basic trends are expected for an electronics: an improvement of a construction (an improvement of strength characteristics for a load-carrying construction, an element of body, *etc.*), a functionality (an improvement of already applied and application of new properties using a nanotechnology), and their combination. The already known nanodimensional coating and nanoparticle is applied to increase the material strength characteristics by several times and factors of a magnitude. We know a filtration, the catalytic, and the absorption properties of a nanoporous material. A nanoparticle (Co, Ni), a nanocrystalline iron film (ZrN, AlN), a superlattice like Nb/Fe, Nb/Ge, an ultra dispersion powder has the unique magnetic properties. A wire nanocomposite (like Cu-Nb), a conducting nanostructured film TiN, TiB_2, the nanoparticle of a metal and a polymer, and a nanotube allow a simultaneous improvement of the characteristics of an electric conductivity and a strength. Due to a high Q-factor, a thermalelectric nanomaterial (a superlattice based on the quantum points PbSeTe, a quantum wire SiGe, and a quantum wall $PbTe/Pb_{1-x}Eu_xTe$) are considered to be promising for a solar energy conversion system and in a cryogenic engineering. A material having a high dielectric penetrability is applied as a multilayer capacitor, a temperature-sensitive resistor, a variable resistor, a memory element, a high sensitivity detector, *etc.* A nanostructured metal-dielectric-semiconductor system is a basic element of an IC (an integrated circuit). A transition to a nanosemiconductor is accompanied by a shift of a luminescence spectrum to a short-wave region and an increased width of a prohibited zone. These phenomena can find a very important application in engineering. A single crystalline particle in a polymer matrix is considered as a possible light-emitting diode, an optical switcher, and a sensor. An application of a heterostructure with the quantum wall and the superlattice like AlGaAs/GaAs in a semiconductor laser allows a decrease of a threshold current and a shorter emission wave increasing a speed of a response of an optical-fiber system. The nanoconductor and especially the nanotube seems to be the most promising material for an emitter, a transistor, and a switcher of a new generation [21-24]. Finally, an electro-mechanical nanosystem (*EMNS*) allows an application of a macro-and nano-world in a whole variety of electronic devices. A field of *EMNS* application is a superminiature sensor, an electro-motor, a transformer, a transducer, a gate, a valve, a capacitor, a resonator, a generator, *etc.* It is reported that a displacement occurring at a level of a thousandth fraction of a nanometer is now possible using the *EMNS* constructed on the basis of GaAs sensor (3000 x 250 x 200 nm) with a one-electron transistor [25].

As it is known, any integral scheme is a highly organized composite structure. Therefore, a singlecrystal is considered as the most suitable material for a technical development in the field of a microelectronics. However, in the course of this development, it turned out that a strict internal crystalline order seriously limits the formation of the integral microscheme. In this case, natural reaction seems to change in favor of a planar technology, a wide application of a film structure, and other materials, *i.e.* in fact, to give-up of a volume (three-dimensional) system. As a result, the planar (2D) technology becomes a technological basis for a modern solid-state electronics. In this case, for a further increasing of an element density, either a three-dimensional (3D) nanosystem, or a molecular electronics should be employed [28].

A physical and a chemical method, which is based on the principles of the self-organized nanodevice, allows a volume density of elements in the 3D-system to be from 1×10^{14} to 5×10^{14} cm^{-3}. A transversal element size is from tens to hundred nanometers. In addition, such composite works under such current densities, which are lower by 3 to 4 orders of a magnitude than those of the planar system. Though recently, due to an application of various types of a lithography and an accelerating device, an effort to decrease a size of a semiconductor element is successful, the planar technology comes very close to a physical capacity. Consequently, a turn to the three-dimensional nanosystem (a nanoelectronics) becomes inescapable.

Even now, it is expected that the application of the nanostructural material in an optical electronics, a system for a control and a transition of a light flow, and an optical-fiber system for a communication will serve a base for a new generation of elements. These systems are a high-rate device for an information delivery, a low-threshold laser and an amplifier, an integral and a close-field optics, and an optical computer, and a system aided to record, to treat, and to display an information using optical methods.

9.3. APPLICATION FIELDS FOR MICRO AND NANODIMENSIONAL STRUCTURES FORMED BY FOCUSED BEAMS OF CHARGES PARTICLES

A nanostamping. A projection-light, an X-ray, an electron, or an ion lithography are the most popular processes, which are employed for a mass-production of a small-sized structure of a high density covering large areas. However, an increasing technical complexity and a rising cost, which are expected in the production of structures with a characteristic size lower than 100 nm, predetermine a necessity to search some alternative solutions. In some cases, the lithography with an application of a focused beam of charges particles in a combination with the nanostamping may turn to be useful [29, 30].

Fig. **2** shows a succession of the processes involved in the fabrication of a 3D Ni stamp using a *PBW* (p-beam writing) technology and a nickel plating. First, Si (100) substrate is coated by Cr (20 µm) and Au (200 µm) layer for a better adhesion and electrical conductivity. Then, a layer of a resistive *PMMA* material is deposited using a centrifugal machine and irradiated by a 2MeV focused proton beam (a). Second, the plated coating is deposited to a top surface playing a role of a stamp base and providing a conductivity in the course of the galvanization (b). Third, a three-dimensional structure is formed as a result of treatment by a developer (c).

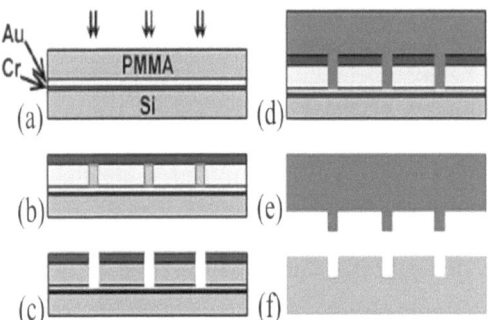

Figure 2: A schematic presentation of a stamp fabrication using a *PBW* technology [27].

Figure 3: A proof of Ni stamp formed in 8 µm layer of a resistive *PMMA* material. An insert shows the stamp under a lower magnification [27].

Fourth, nickel is deposited using the galvanization (d). Fifth, the stamp is separated from the template (e), and the nanostamping is realized (f). Fig. **3** shows a proof of Ni stamp in the *PMMA* layer of 8 µm thickness, which is deposited to Si substrate using the centrifugal machine. You can see pronounced channels of 100nm width and 2 µm depth with smooth vertical walls. Such stamps may be employed up to 15 times without an essential worsening of a reproduction quality [39, 49].

Biomedical applications. With the help of the *PBW* technology, a narrow channel of several scores of a nanometer width, height of which is ten times higher than its width, can be formed. This allows a bimolecular study. Fig. **4** shows a prototype of a biosensor structure representing the comb-shaped electrodes with about 85 nm gap. The biomolecule is studied by a measurement of a full electrical resistance between the electrodes, since every type of the biomolecule has a definite electrical conductivity. The above-mentioned sensor can be applied for a wide variety of biomolecules, including a simple toxin, such as a formaldehyde [31], a big-size *DNA* (a deoxyribonucleic acid), a hormone [32], or a more specified molecule such as an AIDS antibody [33].

Fig. **5** shows another example of 3D microchannels applied for researches of a fibroblast cell behavior aimed at a fabrication of a tissue fiber.

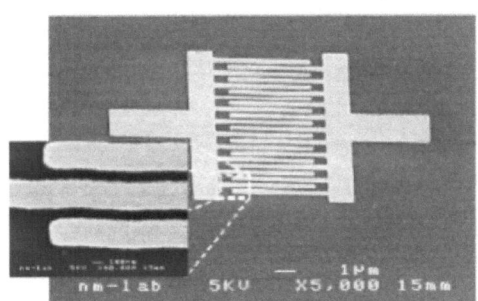

Figure 4: A *SEM* image of a nanobiosensor structure [27].

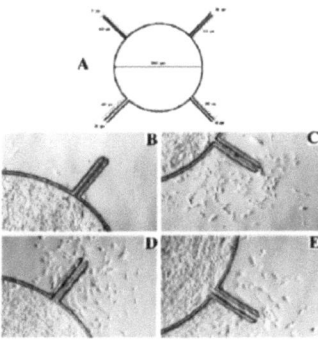

Figure 5: A cell of a fibroblast migrating and penetrating through a narrow channel towards an outside of an incubation region during 7 days. The channels widths: (*B*)-5µm; (*C*)-20µm; (*D*)-25µm; (*E*)-30µm [27].

Here, four channels of 5, 20, 25, and 30 µm width joining an external medium with a closed incubation region are demonstrated. In this Figure, one can see that the cell fails to exit out of this closed region through 5 µm channel. At the same time, the rest channels are suitable enough for its motion.

A microjet system. Another possible application of the narrow micro-and nanochannel, which is formed in a polymer material like the *PMMA* using the *PBW* technology, is a microjet system. Such system can model a liquid flow moving in a tissue of a living organism. Recent ten years, a need of a mass production of the microjet system for fundamental researches performed in the field of chemistry, biology, and medicine essentially increased [34, 35]. However, a nanojet system is now under study, since the channel width lies in a molecular range and its behavior in a flow of a one-molecule range needs further studies. A surface charge of the nanojet system appearing at a nanochannel wall due to an ionization of an OH⁻ group (a hydroxide group) and to a chemical attraction force (Van der Waals force) crucially change a kinetics of an out-flowing liquid. An electro-kinetic over-pumping can be considered as one of the ways to overcome the forces preventing the flow motion through the nanochannel.

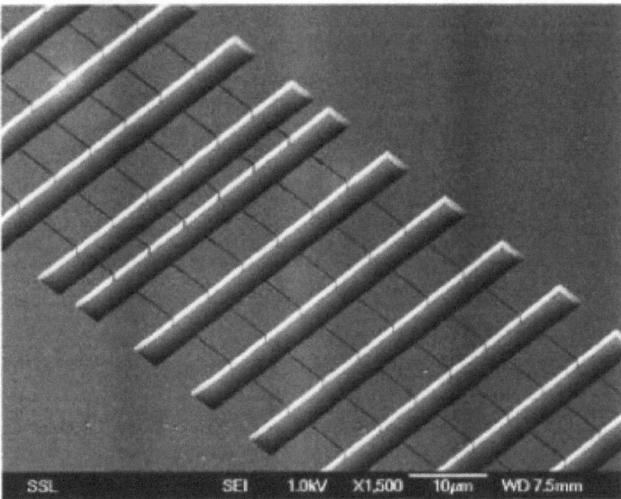

Figure 6: A *SEM* image of reservoirs joined by channels of 100 nm width [26].

Fig. 6 shows reservoirs joined by channels of 100 nm width. When an electric field is applied along the channels, such molecules as a *DNA* rush along the nanochannel. A motion rate depends on a *DNA* size and its scanning characteristics. To some extent, the nanojet system is a nanochip [6]. A roughness value of the channel wall is also important. Since a vicinity effect is practically absent in the PBW technology, a maximum roughness measured by an atomic force microscope reaches < 3 nm [36].

A microphotonics. Among a total variety of applications of the micro-and the nanostructure, which is formed using the *PBW* technology, a microoptics and a microphotonics takes a special place. A light signal is processed using a chip-integrated optical element, such as an emitter, a wave guide, a detector, a modulator, and a bulky of microlenses, which can provide a high rate of information processing.

There are two ways how the *PWB* is applied in the microoptics and the microphotonics. The first one is a direct formation of a small-dimensional structure from a polymer deposited to a corresponding substrate such as a glass or a thermally oxidized silicon plate, using a centrifugal machine. In this case, the substrate and a deposited material should have a lower refraction coefficient than a material of an optical fiber core. A resistive material SU-8 is most acceptable for manufacturing of a wave guide, since it owns a high transparency, low loss, and a smooth wall (Fig. 7) [37]. In addition, its refraction coefficient is a little higher than that of a substrate material made of a usual glass or a thermal oxide. The bulky of microlenses is based on a structure containing a resistive material layer deposited to a glass-like substrate (typical for a light microscopy) using the centrifugal machine. The next step is manufacturing of a mother, which is applied to manufacture the lenses of a desired diameter. After development, the polymer is thermally melted by heating the total structure to a temperature of a glass phase transition. An action of a surface tension forms a semispherical micro-lens (Fig. 8). A focus length depends on a combination of a lens diameter and a thickness of a resistive material. Also, other structures can be manufactured: for example, a lattice or a plate with a Frenkel zone [27]. The second method, which is applied to form the wave guide in a bulky volume of the polymer material or the quartz glass, is a direct ion beam modification without a development stage [38].

It is realized due to processes occurring at a final stage of an ion motion in a sample, when a hidden channel of the waveguide is formed in a substrate. An ion owns a unique feature that an energy amount, which is contributes to the substrate, quickly increases. Therefore, a probability that the ion would form a vacancy at the end of its path increases also quickly. As a summary effect is that a hidden region of damages, which increases a material local density, is formed. As a consequence, a refraction coefficient increases. This damaged region plays a role of a waveguide core.

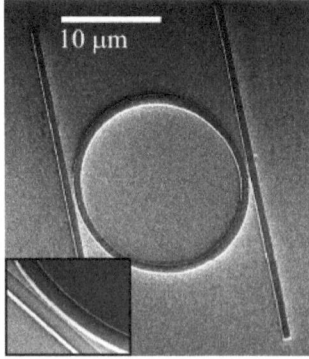
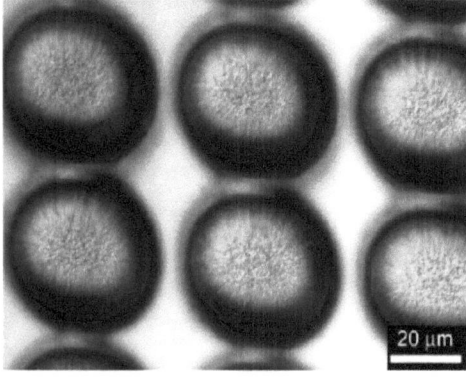

Figure 7: A *SEM* image of a ring vibrator manufactured in a layer of a resistive material SU-8 on a silicon substrate. An upstream image demonstrates a characteristic channel size of 200 nm [27].

Figure 8: An optical image of a microlense bulky manufactured of a resistive *PMMA* material of 15 μm thickness [27].

Among the considered applications of three varieties of the focused charged-particle beam, the *PBW* technology applying a light ion beam has a higher potential in the formation of a low-size structure.

However, a resolution capacity of a nuclear microprobe-a device, which is applied to focus a light ion of an average energy, is now worse than that of other two devices. On the other hand, there are no principle physical limitations for improving the characteristics of a nuclear microprobe. Therefore, a way, how it could be amended, seems to be found in a principally new scheme of the probe formation, a construction of a new lens system, and an application of a new ion source with a brightness, which could essentially exceed that of an already existing high frequency ion source.

9.4. POTENTIAL APPLICATIONS OF CARBON NANOTUBES

Since the time when Prof. Iidjima discovered a carbon nanotube (*CNT*), all researches focused at how to apply it in various devices. In particular, a high success of the *CNT* application is expected from an electron emission due to the above-mentioned *CNT* properties: an extreme physical strength, a chemical stability, a high aspect number, and a high electrical conductivity. Today, the production of a low voltage display is based on a field emission (*FED*). Therefore, the *CNT* is one of the most promising directions of an electronic engineering. To study the field emission devices based on the *CNT*, which were formed by various methods, a single many-wall *CNT* (*MWCNT*) and a bundle of *MWCNT* were employed. In 2004, it was reported about a 32-inch *CNT* based on the *FED* with a subgate type cathode. Developers of this display applied a screen-printing technique allowing a deposition of the *CNT* on a cathode and formed in this way, an emitter. This year, another Japan firm presented a 40-inch triode panel with the cathode coated by the *CNT*. Prof. Pirio fabricated a microstructured cathode with the *CNT* emitter, which was directly formed at a bottom of a cathode hole using a *PECVD*-method at 700°C. Prof. Shiratory formed the field emitter at a glass surface employing a synthesis of a vertically oriented *CNT* at 400°C using the *PECVD* method.

The *CVD* method is also employed to deposit the *CNT* over a large area of an *AAO* template (an anodic aluminum oxide template), which is a prototype of a flat display with the field emission. The *CNT/AAO* is characterized by a low field component of about 2.8V/μm, a high maximum density of an emission current of about 24mA/cm^2, and a good emission stability. Using the *CVD* method without a catalyst in a glowing discharge plasma, a thin-film material is formed from the oriented *SWCNT*. The perfect low voltage electron-field emission properties of the nanotube are notable. An emission current density reaches 50 mA/cm^2 in 5 V/μm field. In comparison with an ordinary emitter, the carbon nanotube demonstrates a lower threshold electric field, see Table **2**.

Table 2: Values of a threshold electric field for different materials under 10mA/cm^2 emission current.

Material	Threshold Electric Field (V/μm)
Mo points	50-100
Si points	50-100
Diamond of *p*-type conductivity	130
Amorphous diamond	20-40
Deposited diamond	20-30
Graphite powder (< 1mm size)	17
Nanostructured	3-5 (unstable under > 30 mA/cm^2)
CNT	1-3 (stable under 1 A/cm^2)

A current density of the carbon nanotube is essentially higher than that of an ordinary emitter. For example, a nanodiamond fails already under lower than 30 A/cm^2 current densities. The *CNT* emitter is attractive for various applications of a vacuum microelectronics and a microwave amplifier of > 500 mA/cm^2 current density.

A cathode-beam illumination element based on the *CNT* were fabricated by Ise Electronic Co., Japan. This illuminating element has a triode-based design. Various colors of its elements were obtained due to an application of different fluorescent materials. A light intensity of a luminescent screen is a factor of two higher (8000 hour lifetime) than that of an ordinary thermal electron-beam tube under the same conditions.

Since the *MWCNT* edge can properly conduct a current, it is applied in a *STM* (a scanning tunneling microscope) and an *AFM* (an atomic force microscope) device and other scanning devices, for example, an electrostatic force microscope. An advantage of the nanotube edge is its flexibility and a capability to scan a thin structure (such as a very small, deep surface crack), which cannot be scanned by a greater, blunter, etched Si or a metallic probe. A biological *DNA* molecule can also be detected by a high resolution head of the nanotube instead of the ordinary *STM* end cap. The *MWCNT* and the *SWCNT* have now such resolution, which could not be earlier reached, and are employed for researches of a biological molecule. Also, one can apply the nanotube edges for a submicron lithography. A *CMP* (a collagen mimetic peptide) looks very promising also for a system for a data storage due to a high wear resistance and can be successfully applied to improve a power efficiency in a device for a thermo-mechanical data storage. Not long ago, a system for the thermo-mechanical data storage on a polymer base (a methylmethacrylate), which employed the *MWCNT* edge, was demonstrated. In this device, an indentation density reached > 40 GigaBit/cm^2.

Recent researches also demonstrated that the *CNT* can be employed as a promising chemical or a biological sensor. As it was found, a specific electrical resistance of the *SWCNT* was essentially changed by an action of a gaseous surrounding medium containing NO_2, NH_3, and O_2 molecule or a biomolecule. It was noticed, that a reaction time of the sensor based on the nanotube was, at least, an order of a magnitude shorter (several seconds were necessary to change a resistance by an order of a magnitude) than for a sensor employing an already available metallic oxide and a polymer. Most often, a chemical sensor is based on a field transistor, in which the carbon nanotube plays a role of a channel.

The *CNT* has many potential applications as a nanochannel for precision delivery of a gas or a liquid. The nanotube features a transportation rate, which is higher by several orders of a magnitude than that of a zeolite or any microporous material. This phenomenal result is related to a smooth inside wall of the nanotube. It is worth mentioning that the carbon nanotube may be efficiently applied as a strengthening component of a composite material, since it features a high strength, a low weight, and a high thermal resistance (for example, its application may be related to sites of a space ship, of an air plane, *etc.*). Not long ago, NASA invested a high amount of money for projects, aimed to form a composite material based on nanotube production technologies, which could be applied in future missions to the Mars. There are also definite advantages in the *CNT* application for a structured polymer compound (for example, an epoxy resin). A material, which is strengthened by the nanotube, increases its strength due to an energy absorption. It is especially important for a ceramic composite matrix based on the nanotube. It is known, that a doping of a small amount of the *CNT* in a polymer composite dramatically enhances a heat conductivity of the polymer matrix. In the course of such treatment, the heat and electric properties of the composite such as a *SWCNT*-based polymer are significantly modified by a magnetic ordering. Joining of nanotubes in the course of the composite formation is an important factor improving the electrical, and especially, the heat transportation properties. Since the nanotube had the relatively direct and narrow channels, immediately after this discovery, it was assumed that these cavities could be filled by a foreign material allowing the formation of a one-diemensional nanowire. In such a way, the nanotube can be employed as a template for the formation of nanowires of various compositions and structures. Also, the nanotube can be filled-in with a metallic or a ceramic material. The nanotube filled-in by SiC, SiO, BN, C can be synthesized *in situ*, in the course of its formation in an electric arc or by a laser ablation. For various purposes, the nanotube may be decorated by a metallic particle and a fullerene. Not long ago, one more interesting application of the nanotube was found. A nanotube was filled-in with a metal Ga, C, or MgO and applied as a nanothermometer. A hollow space inside the SWCNT as a one-dimensional space can be successfully employed for a physical, an electronic, a chemical, and a biological need. Another intriguing possibility is an application of the nanotube incorporated by the fullerene. The nanotube filled-in with the fullerenes (a peapod) were revealed in 1998 by Prof. Smith. A method allowing large-scale formation of such structures was later developed by Prof. Bandov. A spin ordering taking place inside the peapod-system may essentially affect a development of memory devices in future. In our opinion, the *CNT* and the *CNT* bundle, which are filled-in by the fullerene, are of a high interest as an X-ray refractory lens and can be applied for focusing of an emission and a formation of an image. Experiments with an electron transition through the *MWCNT* performed by Dr. Kruger demonstrated that the *CNT* can be applied for focusing of electrons in a nanosize spot. In this experiment, an electron beam was focused in a projected nanotube center. As a result, an electron intensity increased by several times. Also, we should like to add that bundles of the oriented *CNT* are highly interesting for a nanocollimation of a charged particle and a quantum beam. Due to the channeling effect, they can serve as a waveguide, are very promising for a nanolithography and a local analysis of materials within a range of a nanometer resolution. Recently, it was reported that a composite of the *SWCNT*-a polyamide not only featured a decay time lower than 1ps, but also was characterized by a high non-linear third-order polarization. Therefore, this composite seems to be highly promising for an application in a high-quality optical switch.

After experiments, which evidenced that a semiconducting carbon nanotube could function as a field transistor, further progress was reached. An application of the dielectric thin film in a control gate allowed a decrease of a working voltage approximately to 1V. It was also found, that a behavior of a *p*-type transistor based on the *CNT* was a contact phenomenon rather than a bulk property. Today, there are many molecular devices based on the *SWCNT*: various types of a field transistor, a single-electronic transistor working at room temperature, a logical scheme, an inventors, and an electro-magnetic switcher. Recently, it was reported about an experimental model of a flash memory device based on the *SWCNT*, a capacity of which reached 40 GGBit/cm^2. The production of an *n*-type transistor is technologically important, since this allows manufacturing of a complementary logical device and a scheme based on the *CNT*. Experiments demonstrated that a conversion of a *p*-and an *n*-type of a *CNT*-based field transistor may be realized by doping of an alkaline metal to a tube surface or simply by annealing of a transistor device in a vacuum or an inert gas.

9.5. BIONANOTECHNOLOGIES. ARTIFICIAL COLLAGEN

Undoubtedly, the most complicated and most multi-functional nanostructure is a molecular complex regulating and controlling a biological system. Here, a protein should be considered as an example of a nanosized molecular complexe participating practically in all biological processes relating to a molecular transportation, a metabolism, the sensory, and the information properties.

A bionanotechnology deal with a research of properties of a biological nanostructure at a molecular level and with their application. In such a way, this technology is a boundary between a chemistry, a biology, and a physics. Main directions the modern bionanotechnology are a fabrication of an auxiliary biologically compatible material and a development of a technology for its formation:

- Monitoring of a state of an organism at a cell and a subcell level;

- Production of new biological sensors;

- Production of new medications.

The bionanotechnology is not aimed at a fabrication of a biological material such as a protein, or a genetically modified plant, or a living organism with the purpose to gain the new properties.

The main accent of the bionanotechnology is a medicine-a development of a new medications, an improvement of the existing technologies for a diagnostics, a treatment, and a monitoring of an organism state with a super-high resolution (much more higher than in the modern technologies based on a nuclear magnetic resonance).

The bionanotechnology has a high potential allowing a development of the new methods of a purposeful drag delivery, which could help to deliver the drag not simply to a diseased organ but to a concrete sore cell. For this purposes, a functional nanoparticle containing a therapeutic component (medication) and a sensor component are intended. The medication is to be released to a cell, and the sensor component is to determine at what moment it is to be released. Today, an individual stage of whole process has already been demonstrated (determination of a desired cell and a medication release). However, a full stage cycle is still under development.

For example, a non-surgical way of the problem solution was offered in the middle of 1980th [39, 40]. In the course of a directed radiotherapy, a cancer cell in a liver was subjected to a destruction. A microsphere containing a nuclide 89 Y is not a radioactive element, but can be activated by a neutron bombardment up to 90 Y with a half-decay period of 64.1 h. This microsphere is injected to the liver through a liver artery, in which it is captured by a smaller blood vessel blocking the blood delivery to an affected place and providing a direct beam treatment of a malignant cell. The microsphere is now at a clinical disposal in a number of countries, such as Canada, USA, and Australia.

Therefore, the development of the nanosystem, which could help to deliver the medication to an affected place, may become a base for the development of new directions in a medicine-an inside-cell therapy, which means the medical treatment without a surgical intervention into a human body.

A collagen is the mostly abundant protein in a human body. It serves as a natural carcass for a cell and can determine a time and a direction of its growing. New properties and application fields of a synthesized molecule imitating the collagen were discovered at the Institute of NanoBioTechnology of the John Hopkins University (INBT).

Using the nanoparticle, this molecule, which is called as a collagen mimetic peptide (*CMP*), can transfer a detailed information about a malignant tumor, such as a direction of a spreading change in a vessel, can deliver a medication to a cell and improve a blood circulation in an implanted organ. Several useful *CMP* properties are described in the work of Prof. S.Michael Yu., who is an expert in a material science, working at the Whiting School of Engineering, and Prof. Martin Pomper, who is an expert in a radiology and an oncology, working at the School of Medicine. The work was published in the BioMacroMolecules Journal in June 12, 2008.

Fig. **9** shows a photograph of a collagen fiber solution of I-type, which was obtained from a mouse tail tendon after an incubation of a gold nanoparticle. A white arrow indicates a nanoparticle position in the collagen. The photo is made with the help of a scanning electron microscope. The collagen creates a space for a growing cell. However, there are places, where the collagen is forbidden. Prof. Michael reported that the tumor or the blood clot, which result in an insult, are composed of the collagen.

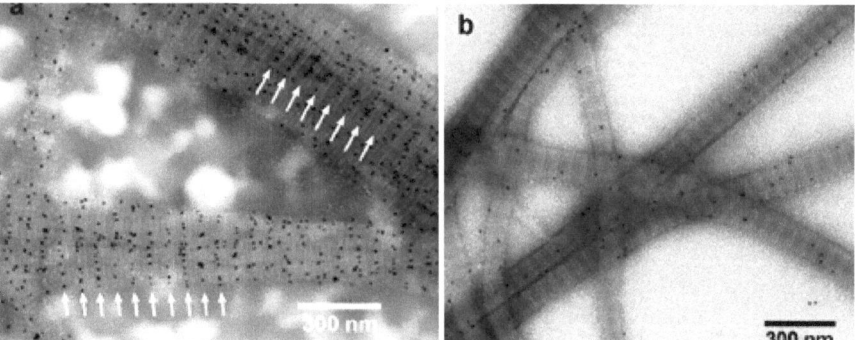

Figure 9: A solution with a collagen fiber.

For the first time, it was revealed experimentally, how the synthesized collagen was bound with a native one in a non-purified *ex vivo* sample (a non-treated tissue). Earlier, only a binding of a purified collagen sample in Petri cap was investigated (the work published by C. Michael Yu). The *CMP* got bound with the collagen accurately laying in its rope-like three-spiral structure. The researches also demonstrated that the *CMP* was built-in into a definite space point positioned along an every collagen chain. C. Michael Yu reported that the *CMP* was able to penetrate imperceptibly into the collagen structure, namely into those space intervals, which were not strongly bound to each other. Using a nanoparticle attached to the *CMP* and a scanning electron microscopy, scientists managed to visualize a concrete region of binding. The region of binding looked like a line containing the black points along the whole collagen fiber length reminding the marks of a usual ruler.

To investigate a *CMP* binding capability in a non-affected tissue, C. Michael Yu. and M. Pomper fixed a fluorescent marker to a *CMP* molecule end. Then, they used a *CMP* solution for a binding with a human liver tissue. As a result, the *CMP* brightly glowed similar to a native collagen, when it is detected by a specific anti-collagen antibody (a usual method applied to identify the collagen). C. Michael Yu. and M. Pomper reported that even in a natural system the *CMP* could be bound only with the collagen and does not affect any other proteins. It has not been never known earlier.

The microscopic nanosized *CMP* molecule is able to create an image of higher resolution than an antibody. The *CMP* is able to penetrate into a region, which is smaller by a factor of 20 than the antibody. Also, M. Pomper presented the images demonstrating an aggregation of the collagen at such points, where it could not be found at all (Fig. **9a, 9b**). Finally, the scientists observed that being heated up to $37^{\circ}C$ (a human body temperature), the *CMP* quickly lost its affinity with the collagen and released. However, a time, during which the *CMP* remains in a bound state with the collagen, can be varied by changing a length of a *CMP* chain. The longer was the chain, the longer was the time, during which the *CMP* remained at its place, as it was reported by C. Michael Yu.

This *CMP* property is applicable to the various therapeutic and diagnostic purposes-a short molecule can be applied for a quick inspection and a longer one-for a longer inspection. A potential *CMP* application is limited only by a molecule type, with which it is planned to be bound. C. Michael Yu and M. Pomper have several patents in the field of therapeutic and diagnostic *CMP* applications. A complex of the *CMP* molecule with a nanoparticle and a nanoshell may be applied for a disease mapping and a medication delivery to a cell.

The synthesized collagen also extends an application sphere of a synthetic implanter. The natural collagen features a chemical signal, which is called a growing factor. It gives an orientation to a growing direction of a cell with respect to a blood vascular. C. Michael Yu applied this factor and the *CMP* for an organized blood vascular net. The work, which was focused at these achievements, was published in a September online issue of the BioMacroMolecules Journal. However, the author is sure that a panorama of this problem is much wider.

The scientist considers that the collagen is a very complicated system, and we do not know everything about it. An application of the synthesized system would allow us to get much information and works perfectly in the case, when we have to solve the mechanical problems about the cell and a skeleton interaction, but scientists still come back to the natural collagen, when they do with the production of an artificial tissue.

9.6. NANOFILTERING AS NEW WAY TO PURIFY DRINKING WATER

A drinking water has a perfect quality, but it is not a reason to be calm. The matter is that a list of new undesirable components contaminating a subsoil water, rivers, and lakes is constantly increasing. The climatic changes result in a change of temperature and a staff of microorganisms in natural ponds. In most countries, a subsoil water often serve as a source of a drinking water, in which one can find an organic contamination. In Swiss, 43% of drinking water is taken from natural springs, 40%-from an underground water, and 17%-from lakes.

The purification is usually required for a water, which is taken from the lakes and the springs. Scientists from WVZ, the Eawag Institute developed methods and a combination of approaches on the water usage under conditions of future changes of its natural composition. Main emphasis is laid on renewing of the water supply and increasing of the equipment lifetime from 30 to 50 years.

The first object was the Lengg Lake. The filters of a small grain sand were changed for an ultrafiltration membrane with 10 nanometer pores (one nanometer is one millionth fraction of a millimeter). In addition to the filtering properties, the membrane filter is an absolute barrier for microorganisms (Fig. **10**). Combined methods of the ultrafiltration, a filtration through an activated coal, and an ozone treatment, each of which can be applied as a self-functioning element, warrant the microbiologically-safe water without an application of a chlorination-a process, which is not popular with consumers. Any trace contaminations can be efficiently removed under such conditions.

Figure 10: An image of aggregated microorganisms.

9.7. SOME INTERESTING RESULTS ON NANOMATERIALS

A review of a scientific literature concerning a modeling of a nanomorphology (that is, a computational and a theoretical examination of a structure, a shape, and a phase of a nanomaterial highlights two common assumptions. An assumption of a crystallographic perfection and an assumption of a chemical isolation. In contrast, a review of a literature reporting about an experimental characterization of a nanomorphology of a real nanomaterial highlights a variety of modifications and defects, and an interdependence between the shape and the chemical environment. In the past, it was assumed in theoretical studies for simplicity that a colloidal nanoparticle was spherical, and that of a nanorod (or a nanowire) was either elliptical or cylindrical. The only difference between the nanowire and the nanotube was that the former was a solid, and the latter was hollow. Recently, advances in the image techniques definitively established a distinct faceting of the crystalline nanostructure. The theoretical and computational studies embraced a variety of polyhedral shapes, and the systematic study of a nanomorphology became more popular. However, if somebody is going to model a realistic nanostructure, which is comparable to that examined by our colleagues experimenters, it is still a long way to go. Firstly, we have to tackle those two assumptions listed above, that limit an opportunity for a direct comparison between a theoretical and an experimental system. The first assumption (of a crystallographic perfection) logically precludes a number of important and valid structures.

Many nanomaterials are naturally defect-free, due to a low diffusion barrier for defects near a surface, but a beautiful variety of twinning configurations observed in a nanometal evidences that the crystallographic modifications may be quite stable. Whether they are a kinetic or a thermodynamic product depends on a method we use to model them but not on the priority they should take in our research.

New methods, which could take into account a crystallographic and a chemical defect, are necessary. The second assumption (a chemical isolation) ignores an important role of a surface chemistry and a surfactant affecting a shape and a surface structure and a moderation of a morphological stability of a sample in storage. The change in a surface chemistry can be either critical to maintaining a desirable shape or structure as changes in temperature.

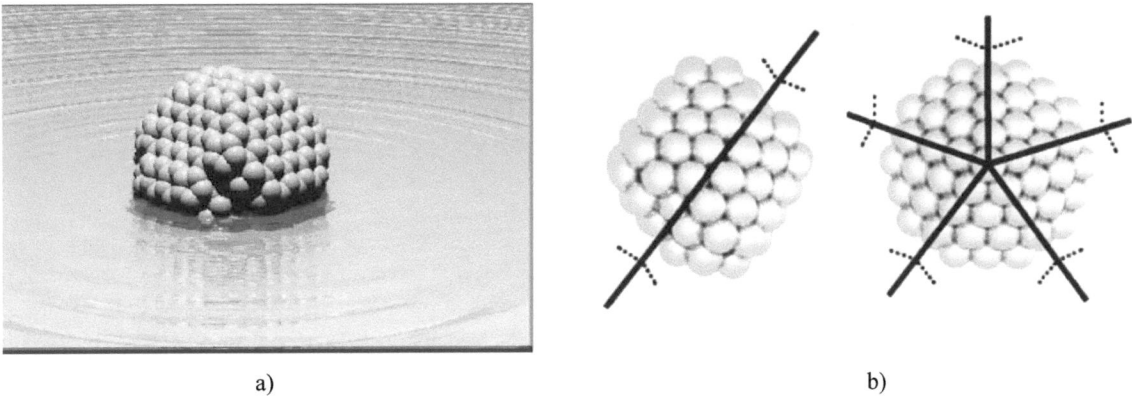

Figure 11: Many colloidal nanostructures are not grown, or stored, under vacuum conditions. It is important to accurately represent the chemistry at the surface if we are to model realistic systems (a) and colloidal gold nanoparticles often exhibit symmetric (contact) or cyclic twinning, which is often neglected in modeling studies that focus on realistic sizes (b).

Moreover, when we are modeling complicated surfactant-particle interactions, perhaps a more realistic basis for comparison should be hydrated or oxidized surfaces, as opposed to the clean surfaces that only stay clean in a virtual environment.

This type of comparison will become more universally accessible with the emergence of new multi-scale methods, which will also facilitate future studies of interactions between the particles themselves, and the modeling of self-assembly and agglomeration [40].

A Drexel University (USA) research team has developed a novel class of biological probes for subcellular investigation [41]. The development of carbon nanotube-tipped pipettes will enable researchers to transfer molecules of interest into and out of an individual cell, nuclei, or organelle through the conjoined nanotube and pipette device. This will allow for controlled substance delivery and quantitative sampling. Davide Mattia and Gulya Korneva, PhD students in Dr. Yury Gogotsi's research team in the Materials Science and Engineering Department, synthesized carbon nanotubes using a template assisted chemical vapor deposition method and filled the tubes with magnetic nanoparticles. Joshua Freedman, a PhD student advised by Drs. Gary Friedman and Adam Fontecchio from Drexel's Electrical and Computer Engineering Department, assembled the probes by injection of mCNTs into glass micropipettes, which are then positioned as probe tips *via* magnetophoresis. and affixed with polymeric adhesive (Figs. **12** and **13**).

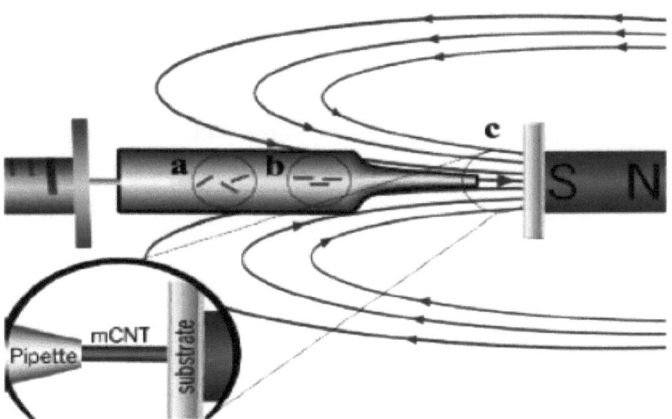

Figure 12: (a) 5 ml of magnetic CNT solution are injected into a pipette using a 30 gauge syringe, (b) As the CNTs approach the magnetic field created by the magnetized wire (right) they align themselves perpendicularly, (c) A thin hydrophilic substrate is placed between the pipette and a powerful magnet. When the pipette is moved within a few microns of the substrate, capillary action pulls the fluid from the pipette until a CNT or CNT bundle is drawn out.

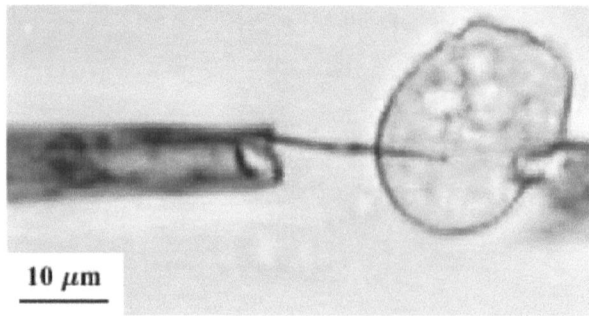

Figure 13: Optical image of a nanotube tipped pipette (left) injected into a MDCK cell held in place by negative pressure on a patch pipette (right). Negligible deformation of the cell occurs.

In their quest to find ways to achieve the much-discussed and sought-after hydrogen economy, the University of Toledo researchers are employing an 'econo' viable and 'enviro' friendly method-well-known to the metallurgists for centuries-to generate high purity hydrogen, with an interesting twist. The method consists of a reaction between heated iron and steam, also appropriately called 'metal-steam reforming'. The twist lies in the source of iron: They utilize the so-called 'mill-scale' waste from steel industry, as an iron source. Mill-scale is a porous, hard and brittle coating of several distinct layers of iron oxides (predominantly Fe_3O_4) formed during the fabrication of steel structures. It is magnetic in nature with iron content up to as high as 93%. Prior to sale, or use of such steel structures, the steel structures must be cleaned of this oxide scale.

Most of the steel mill-scale waste usually end ups in a landfill. In Russia and Asia, some of the mill-scale waste is also used to make reinforced concrete. A purer commercial form of this oxide in combination with nickel and zinc oxide is used in making soft ceramic magnets which are an integral part of all the audio-visual and telecommunication media on this planet as well those in the space.

The mill-scale waste can be and has been successfully converted into metallic iron *via* hydrogen reduction and by reaction with carbon (known as carbothermic reduction) as well. This is not a smart way of producing iron from the oxide; both the reduction processes are energy-intensive as they use high temperatures and one of them requires precious H_2. Thus, the regeneration of elemental iron from the spent oxide *via* these processes is unattractive in a commercial setting and makes the winning of iron from steel waste more expensive than probably dumping it in a landfill! Moreover, the use of high temperatures in the two processes results in the formation of coarser iron that is not so active or 'potent' and is unlikely to generate hydrogen efficiently over several cycles.

So, here is the second element of twist. A novel near room-temperature reduction technique was developed whereby the mill-scale is first brought in solution by acidic dissolution wherein it is instantly converted into highly active nanoscale iron powder (average particle size -20 nm). This new scheme of reduction totally obviates the issue of sintering and coarsening of the iron/iron oxide due to high temperatures and hence the possibility of deactivation during the cyclic operation of metal-steam reforming becomes a non-issue. By modifying the technique and conducting the reduction in the presence of a surfactant, even smaller (-5 nm) iron particles were achieved.

In both these cases an aqueous solution of sodium borohydride ($NaBH_4$) has been used as the reductant: $NaBH_4$ is rather expensive and its solutions unstable. In another case, by using hydrazine as an alternate reductant and ethanol as the solvent under solvothermal conditions (100°C/4atm.), iron particles -5 nm in size have been obtained. This is a significant achievement in that hydrazine is more stable and much less expensive reductant compared to sodium borohydride and. the solvothermal process is easily scalable (Fig. **14**).

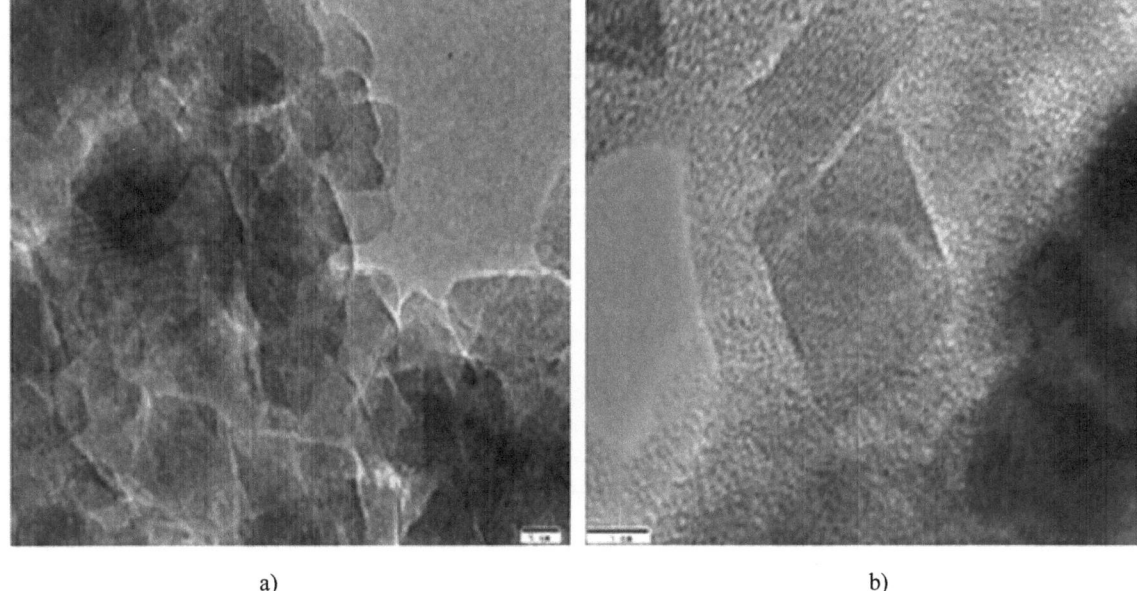

a) b)

Figure 14: TEM images of iron from solvothermal reduction using hydrazine (bar: 5nm).

These zero valent iron (ZVI) nanoparticles are also relevant as key catalyst in the synthesis of carbon nanotubes that are being considered for a host of applications ranging from hydrogen storage devices to sensors to high strength polymer nanocom-posite. owing to their unique hollow structures and exceptional electrical and mechanical attributes. ZVI is being considered as an active decontaminant of drinking water.

Perchlorate is a chemical species of grave health concern due to its interference with the activity of the thyroid gland, and therefore its removal from drinking water sources is very desirable. The improvement in water quality has been negatively affected also by the presence of arsenic in ground water. Severe poisoning can arise from the ingestion of as little as 100 mg of arsenic trioxide.

Chronic effects may result from the accumulation of arsenic compounds in the body at low intake levels. Arsenite (As^{III}) is many times more toxic than arsenate (As^v). The maximum level of arsenic in irrigation water recommended by the Food and Agriculture Organization (FAO) is 0.1 mg/L. The World Health Organization (WHO) recommends that the maximum level of arsenic in drinking water should not exceed 0.01 mg/L (10 ppb). Some countries still accept the level of 0.05 mg/L (50 ppb) in their national standards. Many studies suggest that there is a high possibility of arsenic being taken in by plants from soil or irrigation water, which eventually transfers to humans.

Among several alternatives for arsenic sorption and removal from water, zerovalent iron and its hydrated forms have shown significant propensity of remediation. Some recent research indicates that magnetite (Fe_3O_4) is also effective in arsenic removal from water.

Preliminary experiments conducted at the Lawrence Berkeley National Laboratory, using our 2VI showed that 497 ppb (0.25 mg/L) of arsenic in water could be removed by-1 g of the nano iron powder generated from mill-scale waste, by stirring for 50 minutes [42, 54-57].

The new degree of the freedom of size is indeed fascinating, which not only allows us to tune the physical properties of a specimen but also provides us with new opportunity to derive information that is beyond the scope of traditional approaches. Since the discovery of nanomaterials. there has been great interest in the synthesis and functionalisation aimed towards the development of new functional materials and devices. Compared with the tremendous experimental progress, fundamental understanding of the unusual performance of nanostructures is the key yet remains far from clear.

Recently, Dr. Sun Changqing at Nanyang Technological University, Singapore, published a thematic report dealing with the physical origin of the size induced property change of nanostructures [43] Sun CO. Size dependence of nanostructures: impact of bond order deficiency". By extending the "atomic coordination-atomic size" correlation theory of Pauling [44] and Goldshdmidt to energy domain, Sun proposed and verified the "bond-order-length-strength (BOLS) correlation mechanism that works quite well in predicting the performance of nanostructures in mechanical strength, thermal and chemical stability, acoustic, photonic, electronic, magnetic, dielectric, and transport dynamic behavior of nanostructures.

The size dependence of nanostructures is attributed to the tunable portion of the under-coordinated atoms in the superficial surface skins of at most three atomic layers in combination with the shorter and stronger bonds between the under-coordinated atoms. Atoms in the core interior of the nanostructures remain as they are in the bulk, making no contribution to the size dependency. The broken bonds induced local strain and the associated depression of the potential well of trapping causes the densification and localization of charge, energy and mass, which modify the atomic coherency (the product of bond number and the single bond energy), electroaffinity (separation between the vacuum level and the conduction band edge), work function, and the Hamiltonian of the nanosolid. Therefore, any detectable quantity can be func-tionalized depending on the atomic coherency, electroaffinity. work function, Hamiltonian or their combinations. For instances, the perturbed Hamiltonian determines the entire band structure such as the band-gap expansion, core-level shift, Stokes shift (electron-phonon interaction), and dielectric suppression (electron polarization);

The modified atomic coherency dictates the thermodynamic process of the solid such as self-assembly growth, atomic vibration, phase transition, diffusitivity, sinterbility, chemical reactivity, and thermal stability. The junction effect of atomic coherency and energy density dictates the mechanical strength (surface stress, surface energy, Young's modulus), and compressibility (extensibility, or ductility) of a nanosolid. Most strikingly, a combination of the new freedom of size and the original BOLS correlation has allowed us to gain quantitative information about the single energy levels of an isolated atom and the vibration frequency of an isolated dimer, and the bonding identities in the metallic monatomic chains and in the carbon nanotubes. Further extension of the BOLS correlation and the associated approaches to atomic defects, impurities, liquid surfaces, junction interfaces, and amorphous states and to the temperature and pressure domains would be more interesting, challenging, and rewarding (Fig. 15).

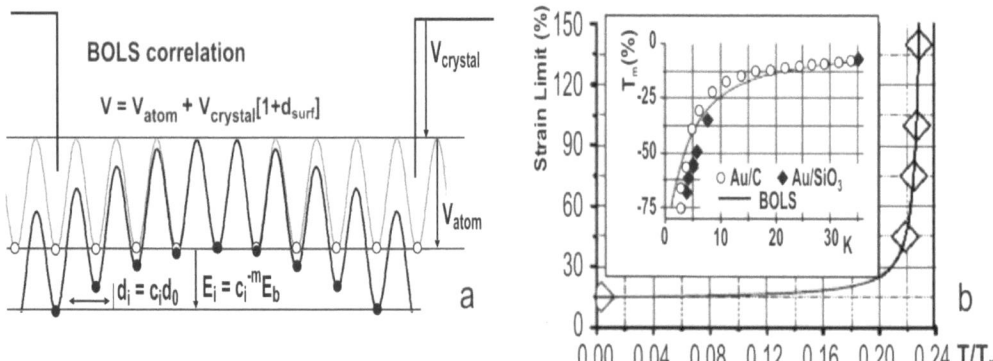

Figure 15: Illustration of the BOLS correlation mechanism and its application, (a) Broken bonds in the surface skin cause the remaining bonds of the under-coordinated atoms to contract spontaneously associated with bond strength gain or potential well depression, which contribute to the Hamiltonian and atomic cohesive energy and electron affinity and related properties of the nanostructures. (b) Agreement between the maximal strain of Au-Au atomic wire at various temperatures and (insert) the temperature dependence of temperature of melting. $K = R/d$ is the number of atoms lined along the radius of a spherical dot. Et and d is the cohesive energy and bond length of the specific i^{th} atom, c is the coefficient of bond contraction, and m is the bond nature indicator.

From a physics point of view, friction is determined by short-and long range interactions between the surfaces. It involves phonon dissipation, bond breaking and formation, strain-induced structural transformation and local surface reconstruction, and adhesion. The classical friction laws were discovered by da Vinci and Amontons and were summarized much later by Coulomb, who contributed the third friction law, stating that friction is independent of the sliding velocity. Recently it has been shown convincingly that ultralow friction can be tailored in nanocomposite materials and that the results point to a breakdown of the Coulomb friction law [45]). Extremely low friction coefficients of the order of 0.01 were attained. It was found that TiC/a-C nanocomposite coatings deposited *via* pulsed-DC magnetron sputtering exhibit a columnar free microstructure that is fully dense. XTEM observations have revealed that the non-reactively sputtered nanocomposite coatings exhibit a multilayer structure. The TiC/a-C top coating consists of amorphous DLC embedded with well aligned Tie nanoparticles of sizes ranging between 2 and 5 nanometers, distributed in a Ti-rich sublayer of the multilayer structure (Fig. **16**). The size and separation of TiC nanoparticles could be monitored independently so as to control the friction coefficient.

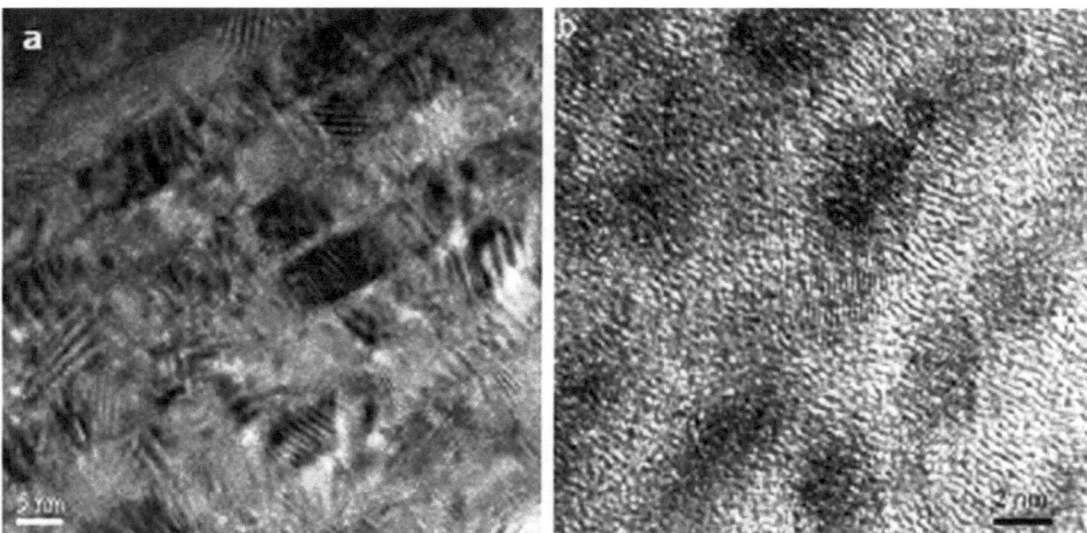

Figure 16: HR-TEM observation of (a) interlayer and (b) top coating.

Researchers at the Biophysics & Nanoscience Centre (BNC), University of Tuscia, Viterbo, Italy have applied spectroscopic and nanoscopic techniques in hybrid systems by conjugating biomolecues with metal nanoparticles and electrodes. They are working towards the development of new strategies to apply in nanomedicine. An emerging task in early diagnostics in medicine is the detection of specific markers at extremely low concentration (less than 10^{-18} M). indeed, abnormal concentration of certain proteins often signals the presence of various cancers and diseases. However, current protein detection methods (for instance ELISA) only allow revealing protein levels above critical threshold concentrations at which diseases are often significantly advanced. The combination of nano-materials with different sensing methods (optical, electrical, electrochemical, magnetic) is currently being explored offering advantageous approaches even for multiplexing sensing.

Growing attention is being shed on advanced optical spectroscopies to obtain detection at very low levels of molecules of interest in biodiagnostics (*e.g.,* fluorescence spectroscopy, Surface Piasmon Resonance, Quartz microbalance, Surface Enhanced Raman Spectroscopy (SERS)). SERS occurring when molecules are adsorbed on nanostructured surfaces of noble metals, couples a high sensitivity with a rewarding chemical specificity representing a powerful microanaiytical technique with large potentialities in advanced nanodiagnostics (Fig. **17**).

Scanning Probe Microscopy (Atomic Force Microscopy (AFM) and Spectroscopy (AFS), Conductive-AFM, Scanning Tunnelling Microscopy (STM)) provide a new, complementary tool to early diagnostics. AFM imaging can reveal individual ligand-receptor complex formation over a substrate as an upward shift of detected height of the molecules. More importantly, AFS allows probing unbinding forces of individual ligand-receptor pairs by recording force *vs.* distance cycles on surface-bound ligands using a tip functionalised with the receptor (see Fig. **18**).

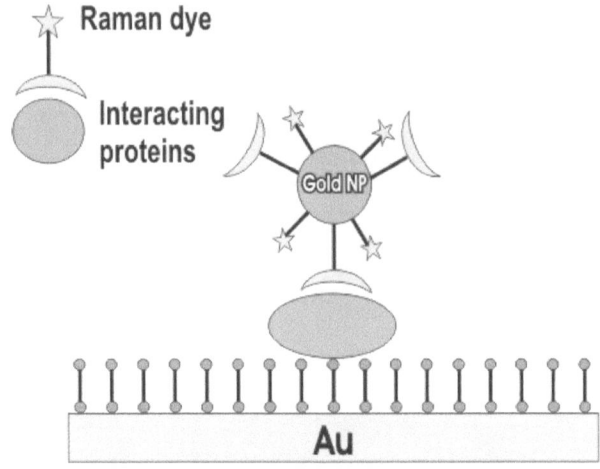 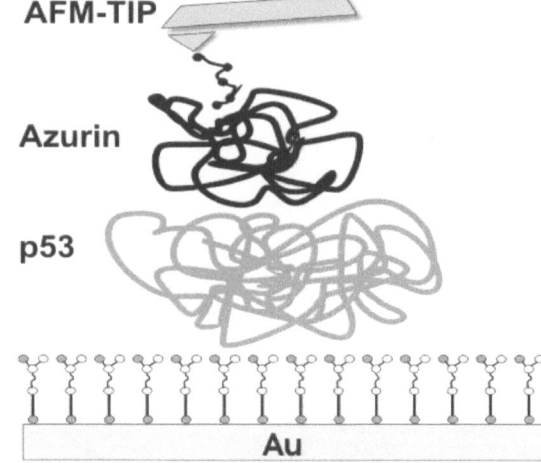

Figure 17: Gold nanoparticles, labelled with both Raman dyes and target molecules, interact with a protein partner adsorbed on gold modified by a self-assembled monolayer. The system can be investigated by spectroscopic (SERS) and nano-scopic (AFM, Conductive-AFM; STM) techniques.

Figure 18: Atomic Force Microscopy tip functionalized with the electron transfer Azurin & p53 immobilized on gold modified by self-assembled monolayer to perform Atomic Force Spectroscopy (AFS) experiments.

Furthermore, the employment of a conductive substrate allows for single specific recognition events to be monitored by STM or Conductive-AFM also as a recordable low current signal (as low as 0.1 pA) with spatial resolution in the nanometer range.

The combination of innovative optical and nanoscopic techniques might deserve new possibilities for nanobiosensing. Therefore, it is important that a quantitative correlation among signals coming from the different single-molecule techniques has to be systematically explored even in terms of quantification, sensitivity, reproducibility and feasibility of detection.

Within this framework, the BNC group are working to develop novel automated nanoscreening platforms which, being based on single-molecule multi-technique detection mode, provide enhanced sensitivity and optimal efficacy of the multiplexing biodetection method, favourably competing with the conventional methods of PCR and ELISA.

Particular interest is being shed on the interaction between p53 tumor-suppressor protein and different ligands. p53 is a major player in regulation of cell growth, DNA repair, genomic stability and cell apoptosis. which is functionally inactivated in many human cancers.

Recent work has demonstrated that, in cancer cells. p53 is stabilized by complex formation with the copper protein Pseudomonas Aeruginosa Azurin (AZ) and that such complex formation results in an increase of p53 intracellular level for ultimate induction of apoptosis and growth arrest of cancer cells. On such a basis, the study of the interaction between p53 and its ligand AZ can provide Insight Into the mechanism of complex formation and specificity of the Interaction, for enhanced p53 activity In cancer cell apoptosis. in addition, the group at BNC are interested to detect Thrombin, a serine protease that converts soluble fibrinogen into insoluble strands of fibrin whose concentration level in blood is relevant in some diseases. As Thrombin sensitive element, the group have used Antithrombin III, forming a complex with Thrombin molecules.

In particular, the BNC group are using gold nanoparticles labelled with Raman dye in connection with gold substrate suitable for performing SERS, AFS and Conductive-AFM experiments (see Figs. **17** and **18**) [46]. Such an approach might offer increased versatility and wide-application as detection method.

Argonne National Laboratories in US provide the worldwide leading fundamental and applied science on the patented UltraNanoCrystailine diamond (UNCD) film technology developed over the last 14 years opening the door to a new era for diamond applications. In general, diamond deposition yields high-performance, long-lasting, radiation-hard dielectric films that can be thin or thick, can be etched alongside silicon components and can only be doped p-type. Diamond's stiffness yields faster resonators, its smoothness yields friction-free microelectromechanical systems and its chemical inertness makes it ideal for bioengineered devices such as human implants. UNCD is not a diamond-like carbon material, but a crystalline diamond film consisting of nanosized grains. Such films, composed of 3-5 nm randomly oriented crystallites surrounded by 0.2-0.3 nm wide grain boundaries are highly electrically insulating unless doped with boron.

A new approach for preparing n-type conductive UNCD films has been developed. This approach entails MWPCVD based synthesis of UNCD from $CH_4/Ar/N_2$ mixture with high nitrogen content. The progressive substitution of nitrogen for argon in the synthesis gas renders the films increasingly electrically conducting with conductivities reaching several hundreds S/cm for 20 % by volume of added N_2. Hall effect measurements have determined the carriers to be n-type with concentrations of 10^{19}-10^{21}/cm^3 and mobilities of several cm^{2}V^{-1}s^{-1}.

The films can be considered insulators in the absence of added N_2 become semiconductors for modest N_2 additions and finally display metal-like behavior at the highest N_2 contents. This enables the controlled variation of synthesis conditions so as to a fabrication of an ambiently conducting n-type diamond film showing the entire range from insulating to semiconducting and even metallic behavior. Moreover the conductivity changes are accompanied by a changeover from strong to weak temperature dependence suggesting that a transition from semiconducting hopping conductivity (0-5 % nitrogen) to metallic conductivity (5-20 % nitrogen) occurs in this regime. Surprisingly, the transition takes place over a variation of plasma nitrogen content of only a few percent. These films are clearly of scientific interest but are also potentially useful since they provide the only currently available n-type diamond material that is electrically conducting at ambient temperatures. One already demonstrated application of UNCD is as part of a highly rectifying diode which is electrically stable even after repeated cycling to 1300K.

A close correlation between the film structure and the electrical conductivity of UNCD films has been observed. Large area, highly dense UNCD nanowires have been successfully synthesized on silicon substrates within substrate temperatures range of 800-900 C starting from 10% of nitrogen in the gas phase. The resultant diamond nanostructures were uniform wires which vary from 5 to 10 nanometers in diameter and between 80-100 nm in length. The transformation from randomly oriented 3-5 nm diamond crystallites to diamond nanowires surrounded by a largely sp2 bonded carbon sheath is followed in detail making use of the High Resolution Transmission Electron Microscope (HRTEM), Electron Energy Loss Spectroscopy (EELS), Scanning Electron Microscopy (SEM) as well as micro-Raman spectroscopy. General views of the samples are displayed in Fig. **19 (I)** and **(II)** corresponding to a SEM image and a low magnification plane view TEM micrograph, respectively. These images show the presence of elongated nanowires (NWs) embedded by a matrix composed by randomly oriented 3-5 nm crystallites of UNCD. The formation of these nanowires starts to appear when the N_2 content in the gas phase reaches about 10% in volume. Fig. **20** corresponds to a HRTEM micrograph of a NW showing the presence of the lattice fringes with a spacing of -0.21 nm which correspond to the d-spacing of the (111) planes of diamond.

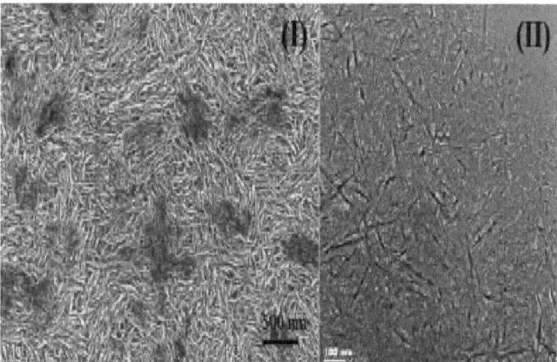

Figure 19: (I) SEAf and (II) low magnification TEM images of the sample showing the presence of NWs.

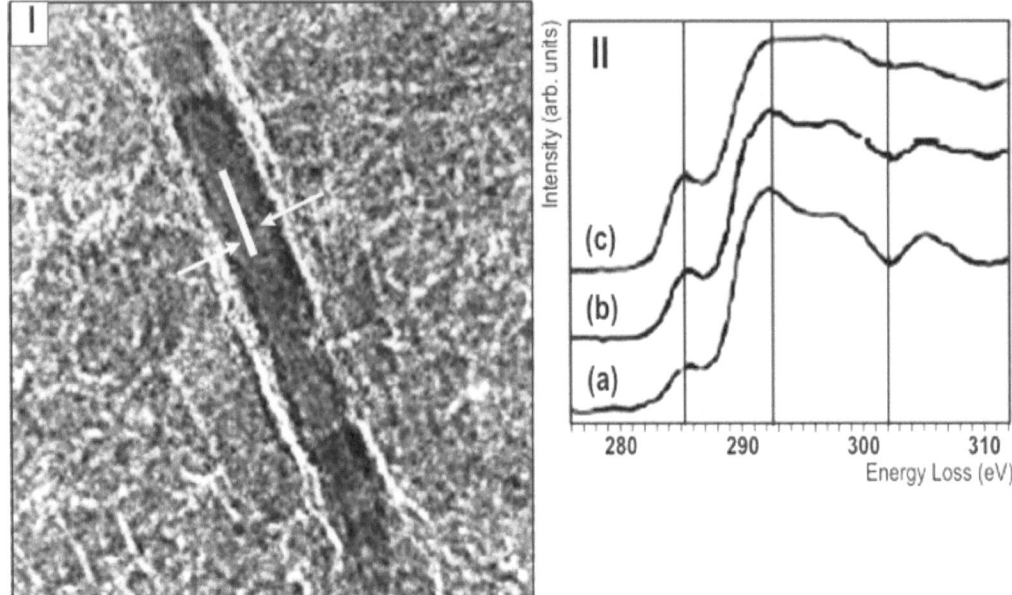

Figure 20: (I) HRTEM image showing one of these Diamond NWs and the matrix. The (111) diamond planes are visible (mark between arrows).

The NWs are enveloped by an amorphous layer of about 1 nm thick. This layer is likely formed during the growth of the NWs. The analysis of the fine structure by EELS confirms that each NW is diamond and that they are enveloped in a sheath of sp2 bonded carbon [47].

From this study it can be concluded that the metal-insulator transition of these films is strongly correlated with the formation of these diamond NWs. These NWs are enveloped by an amorphous carbon layer that seems to provide the conductive path for electrons.

Research Centre of Micro/Nano FabricationTechnology (MNMT) was set up in Tianjin, China in December 2006. This is probably one of the first institutions concentrating on developing micro/nano fabrication technologies for the industry. The research and development work carried out in the centre can be categorised in areas, such as ultra-precision machining of nanometric freeform surfaces. 5-axis machining of brittle materials, focused ion beam (FIB) fabrication for nano features, and micro/nano [48-50].

REFERENCES

[1]   Alfiorov ZhI. Double Heterostructures: Concept of Application in Physcis, Electronics, and Technology. Nobel Award Lecture on Physics.Physis-Uspekhi 2002; 172: 1068-1086.

[2]   Andrievskii RA. Nanostructured Materials-State of Developments and Application. Perspektivnye Materialy 2001; 6: 5-11.

[3]   Munz W-D, Lewis DB, Hosvepian PEh, et al. Industrial scale fabricated superlattice hard PVD coatings. Surf Eng 2001; 17: 15-277.

[4]   Levashov EA, Shtanskii DV. Multifunctional Nanostructured Fuilms. Russ Chem Rev 2007; 76: 502-509.

[5]   Beresnev VM, Pogrebnjak AD, Malikov LV. Structure and tribological behavior of layered TiN-BrAZh 8-4 coatings obtained from metallic plasma flows. J Friction Wear 2008; 29: 35-38.

[6]   Kunchenko YuV, Kunchenko VV, Kartmazov GP. About Increased Resistance of Tools with nanolayered nc-TiNx/CrNx Coatings in the Process of Cutting. Phys Surf Eng 2007; 5: 62-68.

[7]   Beresnev VM, Drobyshevskaia AA. Materialy 8th Mezdunarodnoi Konferentsii "Inzheneria Poverkhnosti I Rennovatsiia Izdelii". Kiev 2008; 27-29.

[8]   Knoteck O, Bohmer M, Leyendecker T. On Structure properties of sputter Ti and Al based hard compound films. J Vac Sci Technol 1986; 4: 2695-2700.

[9] Yao SH, Su YL, Kao WH, Liu TH. On the microdrilling and turning performance of TiN/AlN nano-multilayer films. Mater Sci Eng 2004; 392: 340-347.

[10] Hovsepian PEh, Lewis DV, Munz W-D. Recent progress in large scale fabricationof multilayer/superlattice hard coatings. Surf Coat Technol 2000; 133-134: 166-174.

[11] Ducros C, Benevent V, Sanchette FB. Deposition, characterization and mashinings performance of multilayer PVD coatings on cemented carbide cuttings tools. Surf Coat Technol 2003; 163-164: 681-688.

[12] Loktev YuD. Nanostructured Coatings for High Capacity Tools. Struzhka magazine 2004; 2(5): 12-17.

[13] Beresnev VM, Pogrebnjak AD, Azarenkov NA, et al. Nanocrystalline and Nanocomposite Coatings, Structure, Properties. Phys Surf Eng 2007; 4: 4-27.

[14] Charitidis C, Logothetidis S. Nanomechanical and nanotribological properties of carbon based films. Thin Solid Films 2005; 482: 120-125.

[15] Fang T.-H., Jian S.-R., Chuu D.-S. Nanomechanical properties of TiC, TiN and thin films using scanning probe microscopy and nanoindentation. Appl. Surf. Sci 2004; 228: 365-372

[16] Ed. By Yu S. Karabasov. Novye Materialy. Moscow: MISIS 2002. 736.

[17] Andrievskii RA. Nanomaterials: Concept and Modern Problems. Russ Chem Jour 2002; XLVI: 50-56.

[18] Mukhin VS, Budilov VV, Shekhtman SR, et al. Nanostructured Protective Coatings and Technology of theirfabrication. Kharkov Assembly on Nanotechnologies and Nanoequipment 2006; 1: 205-209.

[19] Belyanin AF, et al. Shock-Resistant Protective Film Coatings Based on AlN in Equipment for Electronics. Teckhnologiia I Konstruirovanie v Elektronnoi Apparature 2004; 4: 35-41.

[20] Malyshevskii VA, Farmokovskii BV. Development of Technology for Supersonic Cold Gas-Dynamical Deposition of nanostructured Functional Coatings for Hydrogen Power Generation. Kharkov Assembly on Nanotechnologies and Nanoequipment 2006; 1: 244-250.

[21] Whitesides G, Grzybowski BA. Self-Assembly at all scales. Science 2000; 295: 2418-2421.

[22] Knobel R, Cleland F. Nanometre-scale displacement sensing using a single electron transistor. Nature 2003; 424: 251-254.

[23] Ruslanov AI. Wondeful World of Nanostructures. J. Gen. Chem 2002; 72: 532-549.

[24] Mistry P, Gomez-Morilla I, Grime GW, et al. New developments in the applications of proton beam writing. Nucl Instr Meth 2005; B237: 188-192.

[25] Watt F, Breese MB, Bettiol A, Van Kan JA. Proton beam writing. Mater today 2007; 10: 20-29.

[26] Jeroen A, Van Kan JA, Bettiol AA, et al. Proton beam writing: a progress review. Int J Nanotech 2004; 1: 464-477.

[27] Ansari K, Van Kan JA, Bettiol AA, Watt F. Fabrication of high aspect ratio 100 nm metallic stamps for nanoimprint lithography using proton beam writing. Appl Phys Lett 2004; 85: 476-478.

[28] Ansari K, Van Kan JA, Bettiol AA, Watt F. Stamps for nanoimprint tithjgraphy fabricated by proton beam writing and nickel electroplating. J Micromech Microeng 2006; 16: 1967-1974.

[29] Dzyadevych SV, Arkhypova VN, Korpan VI, et al. Conductometric for maldehyde sensitive biosensor with specifically adapted analytical characteristics. Anal Chim Acta 2001; 445: 47-55.

[30] Berggren G, Bjarnason B, Johansson G. Capacitive Biosensors. Electroanalysis 2001; 13: 173-180.

[31] Bresler HS, Lenkevich MJ, Murdock JF, et al. ACS Symp. Series, V. 51, American Chemical Society Washington DC: 89-104.

[32] Stephen R. Quake, Axel Scherer From Micro-to Nanofabrication with Soft Materials. Sceince 2000; 290: 1536-1560.

[33] Andersson H, Van den Berg A. Microfluidic devices for celemics: a review. Sensor Actuator 2003; B92: 315-325.

[34] Mahabadi KA, Rodriguer I, Sow Chorng Haur et al. Fabrication of PMMA micro-and nanofluidic channels by proton beam writing: electrocenetic and morphological characterization. J Micromech Microeng 2006; 16: 1170-1180.

[35] Sum TC, Bettiol AA., Van Kan JA, et al. Proton beam writing of low-loss polymer optical waveguides. Appl Phys Lett 2003; 83: 1707-1709.

[36] Sum TC, Bettiol AA, Seng HL, et al. Proton beam writing of passive waveguides in PMMA. Nucl Instr Meth 2003; B210: 266-271.

[37] Kokubo T. Ed. Ben-Nissan, Sher D, Walsh W. Novel Inorganic Materials for Biomedical Applications. Bioceramics 15. Uetikon-Zurich: Trans Tech Publ 2003.

[38] Ehrhardt GJ, Day DE. Therapeutic use of 90Y microspheres. Nucl Med Biol 1987; 14: 233-242.

[39] Wang AY, Foss CA, Leong S. Collagen memetic peptid's binding/specificity to collagen in driven nelical propensity. Bio-macromolecules 2008; 9: 1755-1753.

[40] Hartgerink JD, Beniash E, Stupp SI. Self-Assembly and Mineralization of Peptide-Amphiphile Nanofibers. Science 2001; 21.11: 1684-1688.

[41] Semchiukov YuD. Non-Organic Polymers. Sorosovskii Obrazovatelnyi Journal 1998; 12: 45-51.

[42] Freitas RA. J Exploratory Design in Medical Nanotechnology: A Mechanical Artificial Red Cell, Arti-ficial Cells, Blood Substitutes, and Immobil. Biotech 1998; 26: 411-430.

[43] Drexler KE. Engines of Creation: The Coming Era of Nanotechnology. London: Fourth Estate 1996.

[44] Nanoaffairs 2007; 1 : 2-114

[45] Murugavel P, Lee JH, Lee D, *et al.* Physical properties of miltiferronic hexagounal H_0MnO_3. Appl Phys Lett 2007; 90: 103108.

[46] Changging S. Combination Synthesis of thin mixed oxide films and automated study theit piezoelectric properties. Prog Solid Chem 2007; 35: 1-159.

[47] Amecura H, Ohnuma M, Kishimoto N, *et al.* Fluence-dependent formation of Zr and ZnO nanoparticles by ion implantation and thermal oxidation: an attempt to control nanoparticle size. J Appl Phys 2006; 100: 114309.

[48] Connitraro S//www.units.it/biophysics.

[49] Gruene D//www.anl.gov.

[50] Fang FZ//Asia and Rest of the Word committee.

[51] Hiep HM. A localized surface plasmon resonance based immunosensor for the detection of casein in milk. Sci Technol Adv Mater 2007; 8:331.

[52] Krishnamurthy V, Monfared S, Cornell B. Ion Channel Biosensors Part I Construction Operation and Clinical Studies. IEEE Trans Nanotech 2010; 9 (3): 313-322.

[53] Pickup JC, Zhi ZL, Khan F, Saxl T, Birch DJ. Nanomedicine and its potential in diabetes research and practice. Diabetes Metab Res Rev 2008; 24(8):604-10.

[54] Liao KC, Hogen-Esch T, Richmond FJ, Marcu L, Clifton W, Loeb GE. Percutaneous fiber-optic sensor for chronic glucose monitoring *in vivo*. Biosens Bioelectron 2008; 23(10):1458-65.

[55] Pohanka M, Skladal P, Kroca M Biosensors for biological warfare agent detection. Def Sci J 2007; 57(3):185-93.

[56] Pohanka M, Jun D, Kuca K. Mycotoxin assay using biosensor technology: a review. Drug Chem Toxicol 2007; 30(3):253-61.

[57] Gupta R, Chaudhury NK. Entrapment of biomolecules in sol-gel matrix for applications in biosensors: problems and future prospects. Biosens Bioelectron 2007; 22(11):2387-99.

[58] Pogrebnyak AD, Sobol' OV, Beresnev VM, *et al.* Features of the structural state and mechanical properties of ZrN and Zr(Ti)-Si-N coatings obtained by ion-plasma deposition technique. Tech Phys Lett 2009; 35 (10): 925-928.

[59] Pogrebnjak AD, Lozovan AA, Kirik GV, *et al.* Structure and Properties of nanocomposite, hybrid and polymers coatings. Publ. House URSS, Moscow 2011; 344

[60] Azarenkov NA, Beresnev VM, Pogrebnjak AD, *et al.* Fundamentals of Fabricated Nanostructured Coatings, Nanomaterials and Their Properties, Publ. House URSS, Moscow 2012; 352.

[61] Pogrebnjak AD, Sobol OV, Beresnev VM, *et al.* Phase Composition, Thermal Stability, Physical and Mechanical Properties of Superhard On Base Zr-Ti-Si-N Nanocomposite Coatings Nanostructured Materials and Nanotechnology IV: Ceramic Engineering and Science Proceedings 2010; 31(7): 127-138.

[62] Pogrebnjak AD, Shpak AP, Beresnev VM, *et al.* "Structure and Properties of Nano-and Microcomposite Coating Based on Ti-Si-N/WC-Co-Cr" Acta Physica Polonica A 2011; 120, No. 1: 100-104.

TERMS

This Chapter contains the most important terms and definitions, as well as a list of main abbreviations.

An autoconstruction-is a process of a nanomaterial construction, based on a *"bottom-to-top"* principle, employing a *mechanical synthesis*. The process is performed with the help of a certain automatic system (for example, like a *STM*) functioning according to a given program.

An AFM-is an atomic force microscopy, which is a variation of a *SEM* (a scanning electron microscopy) detecting a force of an interatomic interaction using a sharp mechanical probe.

A domain-is a topologically (in a space) bound region with similar directions of a spontaneous polarization in a *ferroelectric*.

A quantum wire-is a fragment of a conductor or a semiconductor, which is limited in two spatial dimensions and contains the conductivity electrons. The wire cross-section should be so small that a quantum effect could become pronounced.

A quantum point-is a fragment of a conductor or a semiconductor, which is limited in all three spatial dimensions and contains the conductivity electrons. The point should be so small that a quantum effect could become pronounced.

A quantum well is a thin layer of a conductor or a semiconductor (limited in one spatial dimension) containing the conductivity electrons. The layer thickness should be very small that a quantum effect could become pronounced.

A nanocomponent-is a material containing individual nanoobjects of a micro-and a macroscopic size.

A nanocrystalline material is a polycrystalline material, individual crystallites (grains) of which have not less than 100nm size.

A nanobulb-is a carbon nanoparticle, representing the spherical fullerene-like layers embedded one into another.

Nanoscience-is knowledge about properties of matter with nanometer grain range.

A nanoobject-is a matter fragment, which is limited, at least within one space dimension and has not less than 10nm grain size.

A nanoprinting lithography-is a nanotechnological process, which is realized according to a "bottom-to-top" construction principle. It is the formation of a topographic structure with an individual element size now less than 100nm in a polymer surface as a result of a mechanical impression using a special template.

A nanotechnology-is a complex of methods allowing the purposeful formation of a nanoobject with a preliminarily given content, size, and structure.

A self-construction-is a nanomaterial construction according to a "bottom-top" principle based on the self-organized formation of various nanoobjects.

An SOFM-is a scanning optical microscopy of a near-field variation of a *SPM* (a scanning probe microscopy) detecting a near-field optical emission, which is induced by an interaction of a nanometer range light source with a material surface.

A "top-to-bottom"" construction-is a classical principle of the nanomaterial formation involving a "grinding" and a removal of unnecessary parts of a usual microscopic material.

A "bottom-to-top" construction is a principle of the nanomaterial construction involving a construction of a nanomaterial from smaller blocks (an atom and a molecule).

A ferroelectric-is a matter, in which two or more directions of a spontaneous polarization may be realized under conditions of an unavailable electric field. The polarization direction may be changed to opposite by an external electric field. The ferroelectric properties demonstrate themselves in a limited temperature interval, which is lower than a phase transition temperature under a ferroelectric state.

An SPM-is a scanning probe microscopy. It is a wide class of methods based on studies of different material classes with a nanometer space resolution using various probing sensors: for a current, a mechanics, an optics, and *etc*.

An STM-is a scanning tunneling microscopy; it is an ancestor of the *SPM* based on a detection of a tunneling current between a studied material conducting surface and a sharp microscope needle.

A carbon nanotube-is an extended cylindrical structure with a diameter of one to several tens of a nanometer and a several centimeter length. It is composed of one or several hexagonal graphite planes rolled up into a tube, which usually ended by a semispherical head.

A fullerene-is a molecular carbon compound, representing a convex closed polyhedron composed of an add number of three-coordinate carbon atoms, forming a five-and a six-angle ring in a molecule surface.

A fullerite-is crystalline state of *fullerenes,* molecular crystal with fullerenes positioned at lattice sites.

Fulleroid-is a compound based on the *fullerene* formed by joining of a certain functioning group to it.

A chirality-is a property of the *carbon nanotube* arising due to the different directions of nanotube formation from graphite planes. The chirality is characterized by a chirality index.

INDEX

A

AFM (an atomic force microscopy), 84, 86, 90, 128, 137, 138, 143,
Application, 6, 14, 15, 19, 26, 28, 29, 44, 47, 50, 56, 57, 62, 66, 69, 72, 73, 74, 77, 79, 84, 88, 95, 98, 110, 115, 118, 122, 124, 127, 132,
Atom, 5, 9, 11, 15, 17, 19, 21, 24, 25, 27, 31, 36, 41, 48, 51, 60, 65, 74, 77, 84, 86, 90, 96, 101, 113, 116, 123, 126, 136, 138, 143, 144.
Alloys, 16, 22, 27, 28, 47, 51, 53, 63.
Amorphous, 6, 21, 24, 26, 28, 29, 47, 51, 53, 63, 69, 95, 101, 106, 108, 110, 118, 122, 137, 140.
Anode, 58, 59, 66, 72.
Annealing, 24, 27, 53, 70, 109, 110, 119, 129,

B

Beam, 27, 34, 42, 47, 48, 55, 57, 61, 62, 63, 64, 74, 75, 85, 89, 124, 126, 129, 130, 140
Binding energy, 112, 116, 117.
Boundaries, 17, 44, 47, 110, 139,

C

Carbon, 6, 31, 32, 34, 36, 50, 57, 65, 67, 122, 127, 134, 139, 144.
Cathode, 19, 58, 59, 61, 66, 96, 106, 127.
Chamber, 48, 50, 56, 58, 60, 64, 77, 96, 99.
Chemical composition, 3, 23, 33, 55, 57, 62, 63, 65, 88, 118.
Coating, 39, 41, 53, 55, 57, 58, 60, 65, 86, 89, 90, 95, 99, 100, 103, 104, 105, 106, 107, 108, 109, 110, 112, 115, 116, 117, 118, 122, 123, 134, 137.
Coercitive, 10, 12, 28, 41.
Compression, 50, 51, 54, 97, 117.
Curie, 10, 11, 44.
CVD method, 65, 68, 69, 71, 127.
Crystal, 5, 10, 21, 22, 27, 29, 32, 57, 72, 85, 95, 107, 108, 112, 116, 144.

D

Decohesion, 116, 117.
Deformation, 6, 12, 22, 24, 27, 44, 47, 51, 53, 80, 87, 91, 103, 108, 134.
Device, 3, 9, 10, 11, 32, 34, 47, 49, 58, 59, 62, 71, 74, 78, 91, 122, 123, 127, 133, 139.
Diameter, 11, 19, 24, 34, 52, 54, 63, 66, 68, 71, 73, 88, 89, 126, 139, 144.
Discharge, 27, 34, 49, 53, 56, 58, 59, 65, 96, 114, 127.
Diffraction, 23, 64, 84, 88, 101, 117.
Deposition, 19, 27, 33, 41, 47, 53, 55, 58, 60, 63, 64, 72, 95, 99, 107, 118, 139.
Domain, 10, 41, 44, 136, 143.

E

Electrical, 3, 6, 9, 23, 34, 44, 58, 86, 124, 127, 133, 137, 139.
Electric-arc, 48, 53, 62.

F

Fabrication, 19, 21, 26, 28, 33, 39, 42, 47, 50, 51, 52, 55, 67, 73, 95, 113, 122, 129, 140.
FEM(finite element method), 58, 116, 117.
Film, 3, 10, 13, 33, 39, 43, 47, 55, 58, 63, 71, 89, 96, 105, 138.
Friction coefficient, 118, 137.
Fullerene, 31, 33, 41, 50, 65, 69, 128, 143, 144.

G

Gas, 18, 23, 32, 47, 48, 50, 56, 69, 70, 85, 96, 98, 113, 129, 139.
Glow discharge plasma, 58, 62, 65.

H

Hardness, 6, 7, 13, 33, 52, 60, 89, 91, 102, 104, 111, 112, 115, 117, 118, 122.
High-temperature, 18, 88, 112.

I

Indentation , 6, 89, 90, 93, 112, 115, 118, 128.
Interface, 3, 4, 47, 53, 60, 64, 84, 99, 103, 108, 111, 136.
Interaction, 6, 15, 17, 22, 26, 32, 41, 50, 57, 63, 71, 74, 75, 86, 90, 92, 131, 133, 138, 143, 144.
Ion implantation, 53, 60, 62, 63.

J

Jet, 32, 49, 52, 63, 74.
Junction, 4, 7, 32, 66, 108, 136.

L

Laser, 27, 31, 33, 48, 53, 57, 58, 63, 67, 70, 76, 86, 124, 128.
Layer-by-layer, 65.
Liquid, 17, 18, 21, 26, 49, 52, 66, 125, 128, 136.
Loading, 62, 90, 91, 107.

M

Material, 3, 5, 6, 15, 19, 21, 27, 39, 44, 47, 54, 60, 72, 85, 89, 92, 95, 103, 109, 122, 129, 135, 140.
Magnetron sputtering, 58, 59, 65, 96, 103, 106, 109, 114, 119, 137.
Mechanical, 3, 4, 6, 14, 23, 27, 39, 45, 52, 60, 65, 74, 84, 89, 106, 108, 115, 122, 131, 143.
Methods, 16, 21, 26, 29, 33, 47, 49, 53, 57, 64, 72, 78, 84, 88, 95, 118, 124, 127, 137, 143, 144.
Metals, 9, 16, 22, 28, 41, 49, 116, 137.
Modulus, 14, 27, 84, 89, 91, 117.
Magnetic, 3, 10, 12, 21, 26, 28, 39, 44, 49, 57, 77, 84, 89, 96, 122, 128, 133, 136, 137.
Monolayer, 43, 105, 112, 116, 138.
Multifunctional, 118.

N

Nanocomposite, 6, 10, 13, 15, 39, 95, 100, 102, 105, 106, 110, 137.
Nanoobjects, 74, 84, 143, 154.
Nanograin, 105, 106, 118.
Nanotube, 6, 31, 34, 47, 65, 68, 70, 123, 127, 133, 136, 155.
Nuclear scanning microprobe, 74.

O

Oxidation, 44, 70, 102, 112, 118.

P

Particle, 9, 10, 19, 22, 29, 39, 41, 44, 47, 67, 84, 115, 124, 134.
Phase, 3, 6, 9, 10, 15, 18, 33, 39, 43, 48, 55, 61, 64, 92, 95, 100, 105, 109, 114, 118, 143, 150.
Plasma, 27, 32, 41, 48, 50, 59, 96, 99.
Plastic deformation, 6, 24, 25, 47, 54, 92, 102, 116.
Polymer, 18, 19, 28, 39, 41, 72, 122, 135.

Properties, 3, 5, 6, 9, 13, 19, 21, 26, 33, 39, 41, 60, 65, 84, 91, 95, 97, 118, 123, 144.
Pressure, 18, 33, 47, 51, 54, 69, 77, 96, 98, 103, 109, 134, 136.
PVD (a Physical Vapor Deposition), 55, 60, 65.

R

Resistance, 6, 8, 10, 13, 27, 39, 41, 51, 60, 87, 91, 121, 128.
Resistive, 28, 55, 58, 74, 75, 78, 79, 124, 126.

S

Scanning electron microscopy, 74, 75, 84, 114, 130, 139, 143.
Solid solution, 98, 100.
SIMS (Secondary ion-mass spectrometry method), 84, 88.
Single-wall carbon nanotube, 34, 35, 66, 69.
Size, 3, 4, 5, 6, 19, 33, 39, 49, 68, 76, 79, 102, 130, 136, 143.
Steel, 8, 60, 116, 117, 122, 134.
Substrate, 16, 19, 27, 33, 36, 41, 48, 51, 53, 55, 61, 71, 74, 85, 89, 96, 100, 109, 117, 124, 134, 139.
Superconductor, 33, 88.
Surface, 3, 4, 10, 27, 52, 59, 74, 114, 121, 138, 144.

T

Target, 32, 53, 61, 62, 67, 77, 96, 119, 138.
Thermodynamical, 9, 21, 22, 63, 95, 108.
Transformation, 9, 16, 19, 23, 56, 101, 116, 136, 139.
Tensile stress, 113, 116, 117.

V

Vacuum, 19, 32, 33, 36, 48, 51, 57, 60, 65, 77, 86, 127, 129, 133, 136.
Vapor, 31, 41, 48, 49, 55, 64, 73, 119, 133.
Voltage, 6, 19, 57, 58, 59, 66, 77, 86, 96, 99, 129.

W

Wear, 8, 41, 57, 87, 92, 118, 128.
Whisker, 40, 117.

X

X-ray, 23, 59, 76, 87, 95, 100, 117, 124, 128.

www.ingramcontent.com/pod-product-compliance
Lightning Source LLC
Chambersburg PA
CBHW051017180526
45172CB00002B/383